ENVIRONMENTAL SCIENCE, ENGINEERING AND TECHNOLOGY

SOIL ORGANIC MATTER AND ITS INTERACTIONS WITH METALS: PROCESSES, FACTORS, ECOLOGICAL SIGNIFICANCE

ENVIRONMENTAL SCIENCE, ENGINEERING AND TECHNOLOGY

Additional books in this series can be found on Nova's website under the Series tab.

Additional E-books in this series can be found on Nova's website under the E-books tab.

ENVIRONMENTAL SCIENCE, ENGINEERING AND TECHNOLOGY

SOIL ORGANIC MATTER AND ITS INTERACTIONS WITH METALS: PROCESSES, FACTORS, ECOLOGICAL SIGNIFICANCE

GALINA V. MOTUZOVA
IVAN P. MAKARYCHEV
HASSAN M. DERGHAM
ANDREJ A. STEPANOV
AND
NATALIA U. BARSOVA
EDITORS

Nova Science Publishers, Inc.
New York

For permission to use material from this book please contact us:
Telephone 631-231-7269; Fax 631-231-8175
Web Site: http://www.novapublishers.com

Additional color graphics may be available in the e-book version of this book.

Library of Congress Cataloging-in-Publication Data

Soil organic matter and their interactions with metals : processes, factors, ecological significance / editors, Galina V. Motuzova ... [et al.].
p. cm.
Includes bibliographical references and index.
ISBN 978-1-61470-947-3 (hardcover)
1. Soils--Heavy metal content. 2. Soils--Humic acid content. 3. Soil corrosion. I. Motuzova, G. V. (Galina Vasil evna)
TD879.H4S65 2011
577.5'72753--dc23

2011030131

Published by Nova Science Publishers, Inc. + New York

CONTENTS

PREFACE

The soil organic matter is the most important component of ecosystem. As a product of transforming the organic residues under specific conditions it enable to reflect the impact rendered by these conditions. To be able to perform unique functions in the ecosystem the organic matter becomes very active in interaction with mineral components. Among their products the organo-mineral compounds with metal ions are very significant. Being widespread in the environment these compounds play a great role in soil genesis, plant nutrition, migration and accumulation of chemical substances. At present, when the load of heavy metals is augmenting on soil, the consequences of the formation of these compounds in soil require special attention. Especially in the last decade the organo-mineral interactions were the subject of more thorough studies thanks to the development of new profound methods. New experimental data are vital to provide information about scantily studied soils and complexing processes affected by different factors. However, the impact rendered by complexing processes of humic acids with ions of heavy metals upon the properties of humic acids, the humus state and the main chemical properties of soils have been so far examined insufficiently. The aim of these investigations is to receive the answers on the above named questions.

INTRODUCTION

INTERACTIONS OF THE ORGANIC MATTER WITH HEAVY METALS IN SOILS AND ITS ENVIRONMENTAL IMPORTANCE

The soil organic matter is the most important component of ecosystem. As a product of transforming the grass residues under specific (zonal, local) conditions of their humification, the organic matter and its properties in particular enable to reflect the impact rendered by these conditions. Moreover, the humic acid properties have an influence upon soil processes and the interaction of soils with adjacent media. In the ecosystem the organic matter performs such important functions as accumulative, transport, protection, memory, physiological, trophic ones being capable to regulate the soil stability to different kinds of external effects (Orlov, 1990; Dergatcheva, 1997; Popov, 2004). To be able to perform these functions, the organic matter becomes very active in interaction with mineral components. Among interaction products the organo-mineral compounds with metal ions are significant. Being widespread in the environment these compounds play a great role in soil genesis, plant nutrition, migration and accumulation of chemical substances.

At present, the ever increasing degradation of soils due to anthropogenic (agricultural, technogenic) factors, when the load of heavy metals is augmenting on soil, the consequences of the formation of these compounds in soil require special attention. The chemical state of metals proper and humus substances are highly affected by interactions of the soil organic matter especially that of humic origin with metal ions, thus changing the main chemical properties of soils. In the last fifty years there have been considerable efforts to study organo-mineral compounds with metal ions in soil (Bremner et al., 1946; Heinzke and Mann, 1946; Martell and Calvin, 1952; Beckwith, 1955; Broadbent and Ott, 1957). In Russia this study was started by Kaurichev, Khan, Karpukhin, and Fokin (Kaurichev and Nozdrunova, 1961; Kaurichev et al., 1964; Khan, 1969; Fokin and Karpukhin, 1972; Fokin, 1975) in the 1960-70s. Especially in the last decade the organo-mineral interactions were the subject of more thorough studies thanks to the development of profound methods like chemical, physical-chemical (ion exchange, kinetic, spectrophotometry, isotopic exchange), electrochemical (potentiometric and conductometric titration, polarographic), chromatographic, etc. Today, the chemical nature of interaction between metals and organic ligands became evident. The peculiar composition and properties of complexes composing of humic substances and metal ions have been identified as well. The influence of organo-mineral interactions in soil upon the adsorbing capacity of soil minerals in terms of metal ions and their migration in different soils has been profoundly studied. New experimental data are vital to provide information about scantily studied soils and complexing processes affected by different factors.

However, the impact rendered by complexing processes of humic acids with ions of heavy metals upon the properties of humic acids, the humus state and the main chemical properties of soils have been so far examined insufficiently. It is possible to find some answers in the present book.

Using several experimental methods it was established that the complexing of humic acids with ions of heavy metals is accompanied by destroying the composition and structure of natural organo-mineral compounds. Mechanisms responsible for these processes serve as evidence that the links between aromatic and aliphatic components of humic acids become weakened, the content of metals in the composition of compounds with the organic matter is increased, the ratio between humic acids and fulvoacids in humus shows changes. The main acidic properties of the liquid phase in soils contaminated by heavy metals are also changed to a considerable extent. The experimental data showed that the complexing of water-soluble organic substances with ions of heavy metals leads to acidification of the liquid phase in soils. The sources of additional soil acidity, mechanisms of processes providing this effect, quantitative features and affecting factors are also presented in this book.

The experimental data are valid to show that in contaminated soils the interaction between the organic matter and heavy metals causes the humus reorganization and changes in the main chemical properties of soils. Such conclusions are of great importance for the development of theoretical problems in soil science, agrochemistry, biochemistry, environment management, ecology as well as for practical decision-making in the sphere of environment protection.

Chapter 1

HUMIC SUBSTANCES: ORIGIN, TRANSFORMATION, INTERACTION WITH METALS, AND ECOLOGICAL SIGNIFICANCE (REVIEW)

Humic substances are the products of the living activity of organisms on the planet and a condition of its continuation. The process of humus formation is comparable in its scope with photosynthesis. About $50 \cdot 10^9$ t of atmospheric carbon is annually fixed by living organisms (predominantly, by plants) through photosynthesis; after their death, organic residues containing about $40 \cdot 10^9$ t of carbon remain on the earth surface. These residues are partially mineralized to CO_2 and H_2O, and a significant part of them containing 0.6–$2.5 \cdot 10^9$ t of carbon is transformed into humic substances (Perminova and Hatfield, 2005).

The major sources of soil humus are plant residues decomposing under the action of microorganisms. The main components of plant residues are carbohydrates and arenes (lignin, tannins, flavonoids) (Aleksandrova, 1980; Grishina, 1986). Carbohydrates constitute 40–85% of plant residues (on the ash-free dry weight). Among them, cellulose and hemicellulose predominate. The contents of proteins, lipids, and tannins in plant residues are usually low. The chemical composition of different organs in woody plants significantly varies: proteins compose about 8–12% in needles and leaves and no more than 1–2% in woody tissues. An increased portion of nitrogen-containing substances is found in grasses and especially legumes. The direct contribution of microorganisms to the composition of the forming humus is low, but they play the crucial role in the transformation of plant residues. Humus consists of organic substances that have lost their anatomical relations with cellular and tissue structures of plant residues (Orlov, 1990).

Components of organic residues vary in their biothermodynamic stability. Polysaccharides, proteins, and nucleic acids are instable in the soil; they are rapidly consumed by microorganisms. The stability increases in the series: waxes < resins = tannins < melanins = lignins (Saiz-Jimenez, 1996).

Lignin is an important source of humus; its large amounts are contained in plants residues. It represents an irregular polymer with branched macromolecules composed of substituted phenolic alcohol residues and is relatively stable in the soil. The chemical structure of lignin allows the appearance of a large number of different products of its decomposition, including both low- and high-molecular-weight compounds.

Flavanoids (including tannins), along with lignins, are important sources of aromatic structures in soil humus. Their content in plant residues is lower than that of lignin, but they have a high reactivity (Orlov, 1990). The group of lipids includes fatty acids, styrenes, terpenes, hydrocarbons, chlorophyll, fats, waxes, and resins. They compose 2–6% of soil organic matter (Stevenson, 1994). The most important nitrogen-containing substances as precursors of humus acids are proteins, polypeptides, amino acids, amino sugars, nucleic acids, chlorophyll, and some other substances (Aleksandrova and Nazarova, 1978; Orlov, 1990).

There is no consensus on the mechanisms of transformation of organic residues and humification. Two radically different ways for the formation of humus acids are supposed: abiotic condensation and degradation of biopolymers. According to one hypothesis, organic residues are completely decomposed to monomers that are then condensed and polymerized (Kononova, 1963; Flaig, 1964; Flaig et al., 1975). According to the other hypothesis, humic substances are formed at any decomposition stage of lignin, proteins, polysaccharides, and other sources without their complete decomposition to monomers. The major process of humification is the formation of acid functional groups under the impact of oxidation reactions (Aleksandrova, 1980).

Orlov (1990) developed a general theory of humification, according to which humification processes are aimed at selecting biothermodynamically stable products of transformation, regardless of pedogenesis factors and soil type. This explains the common structural principles of humic substances formed in different soils and other natural environments. According to Orlov, the depth of humification is a function of the total volume of organic residues getting to the soil, the rate of their transformation, and the time. The rate of humification depends on the factors affecting the vital activity of soil biota (temperature and water regimes, reaction, nutrient supply of organisms, the presence of toxins or stimulators) and the resilience of plant residues to the biochemical transformation.

Humus consists of nonspecific organic substances, specific humic substances, and various intermediate products of the decay and humification. Specific humic substances (HSs) are classified according to their solubility in different solvents. Humic acids (HAs) are soluble in alkalis and insoluble in acids (pH < 2); fulvic acids (FAs) are soluble in both alkalis and acids; himatomelanic acids represent the fraction of HAs dissolved in the ethanol solution. Humin consists of organic compounds remaining in the soil after extraction of HAs with acids, alkalis, and organic solvents (Stevenson, 1994). Humic substances differ in the composition of the sources of humus and in the conditions of their transformation. However, they are relatively similar in their elemental composition: C, 51–58%; H, 4–6%; O, 32–38%; N, 3–5% (% on the ash-free dry weight) (Tables 1.1, 1.2). Humic substances also contain some amounts of S, P, and ash elements.

Table 1.1. Average elemental composition of HAs of different origins, % on the ash-free dry weight (Orlov, 1990; Perminova, 2000)

Preparation	C	H	O	N
Soddy-podzolic soil HAs	53.4	4.8	37.7	4.1
Chernozem HAs	57.9	4.0	36.7	4.0
Sierozem HAs	56.0	4.8	34.4	4.8
Plant residue HAs	56.1	5.5	33.5	4.9
Peat HAs	57.1	5.0	35.2	2.8
Fresh water HAs	51.2	4.7	40.4	2.6
Marine sediment HAs	56.3	5.8	31.7	3.8

Table 1.2. Average elemental composition of FAs of different origin, % on the ash-free dry weight (Orlov, 1990; Perminova, 2000)

Preparation	C	H	O	N
Soddy-podzolic soil FAs	45.9	4.5	46.4	3.2
Chernozem FAs	43.0	4.7	48.8	3.6
Soddy-podzolic soil FAs	54.2	5.3	37.8	2.0
Chernozem FAs	46.7	4.2	45.9	2.3
Soddy-podzolic soil FAs	45.0	5.9	45.1	4.1

Humic substances are characterized by significant heterogeneity and complex structure. These are mixtures of natural organic compounds formed during the selection of the most stable structures, which are stochastic in character (Orlov, 1990; McCarthy, 2001). Each of their parameters can be described by a distribution function. The basic properties of HSs are nonstoichiometric composition, irregular structure, heterogeneous structural elements, and polydispersion (Beyer, 1996; Perminova, 2000). The real composition and properties of soil HSs in different natural zones are controlled by the combination of biotic (the amount and qualitative composition of organic residues and the number and activity of organisms processing them) and abiotic (solar insolation, temperature, and moisture) factors.

The concepts of HS structure are based on the destructive and non-destructive (spectroscopic) methods of study. The destructive methods include hydrolysis (a relatively mild treatment) and hard destruction (oxidation and reduction). Hydrolysis mainly affects the readily hydrolyzable peripheral part of HS molecules. Hard methods are used for studying the hardly hydrolyzable nuclei of HS molecules.

The hydrolyzable part of HA molecules makes up 30–50%. HAs of soddy-podzolic soils and sierozems are more subjected to hydrolysis than HAs of chernozems, brown coals, and peats. The hydrolysable part of FA molecules is larger than that of HAs; it reaches 60–80%. The composition of hydrolyzate is indicative of structural fragments composing the periphery of HS molecules. Hydrolyzates mainly consist of carbohydrates (65–80%), amino acids (14–17%), and amino sugars (4–6%). Benzene polycarboxylic acids and phenoxy acids (mainly tri- and tetracarboxylic acids) are the major products of the hard decomposition of HSs.

The study of HAs by different methods showed that the HS molecules consist of an aromatic nuclei or skeleton (nonhydrolyzable part) and a carbohydrate–peptide complex (hydrolyzable periphery) (Orlov, 1990).

HS molecules are enriched with functional groups, which determine the reactivity of HSs. The molecules of HSs include oxygen-, nitrogen-, and sulfur-containing groups: carboxyl, phenolhydroxyl, alcohol, methoxyl, chinoid, ether, ketone, amide, amine, imide, sulfo, thiol, and disulfide groups (Orlov, 1990; Perminova, 2000). The carboxyl and phenolhydroxyl groups are most abundant. They determine the capacity of HSs for ion exchange and complexation with metal ions. The carboxyl groups contain 18–65% of total oxygen in HS molecules, and the phenolhydroxyl groups contain 9–38% of total oxygen (Perminova, 2000).

The polydispersity of HSs implies that they are characterized by a set of molecular masses (MMs), i.e., a molecular-mass distribution, from which the average MM value is calculated. The MM value determines the solubility of HSs, their migration ability, their sorption by the mineral soil components, and consumability by living organisms (Orlov, 1990). According to the chemical analysis of HSs, the minimum MM of FAs cannot be lower than 350–450 Da. From gel-filtration data, HAs have molecular masses of 10–200 kDa (Perminova, 2000).

Interaction of HSs with Inorganic Soil Components

HAs are capable to actively interact with different mineral components of soils: cations, oxides and hydroxides, silicates, and aluminosilicates. Organomineral compounds are formed during different processes: weak adsorption interactions of organic and mineral components, chemisorption phenomena, the immobilization of organic compounds in the interlayer space of clay minerals, the formation of salts and complexes, and mutual coagulation and coprecipitation.

HSs can interact with minerals through the binding of HS carboxyl groups to the OH groups of the mineral surface by mineral bridges; under acid conditions, amides are converted to the cationic form, which enables sorption on negatively charged clay minerals due to electrostatic forces (Murphy et al., 1992; Vermeer and Koopal, 1998).

Complexation of HSs with Metals

The complex composition of HSs and the permanent presence of aromatic fragments and carboxyl, hydroxyl, and carbonyl functional groups determine their capacity to enter into various (ionic, donor–acceptor, and hydrophobic) interactions between organic substances and other compounds. The formation of complexes with different chemical elements is of greatest importance. Both natural and technogenic complexes with metals are very important. The complexation of HSs with metal ions significantly affects the evolution of pedogenesis, the accumulation and migration of metals and organic substances, and the nutrition of plants, as well as the availability of metals to plants and the harmful effect of their excess on living organisms. Metals have different affinities for separate functional groups. Such cations as Cd^{2+}, Pb^{2+}, and Hg^{2+} predominantly form complexes with nitrogen- and sulfur-containing groups. The Ca^{2+} and Mg^{2+} cations form instable complexes with oxygen-containing functional groups. Most heavy metals, including Fe^{3+}, Cu^{2+}, Zn^{2+}, etc., can form complexes with different functional groups (Senesi and Loffredo, 2005).

Complexes of HSs with metal ions can have different stoichiometry. One binding site of HS molecule can contain several functional groups and form a chelate with the metal ion. The presence of salisylic- and phthalic-type binding sites in HS molecules is suggested (Senesi and Loffredo, 2005). One metal ion can form complexes with two and more HS molecules, and one HS molecule can add several metal ions. Carboxyls and phenolhydroxyls are the most abundant functional groups responsible for the complexing capacity of HSs. Nitrogen- and sulfur-containing groups present in small amounts in HS molecules are also important, because they form strong bonds with many metal ions (Croue et al., 2003; Tipping, 2005).

Complexes of HSs with metal ions differ in stability. Two main types of complexes are distinguished: inner-sphere and outer-sphere complexes. The outer-sphere complexes involve electrostatic forces; water molecules occur between the metal ion and the ligands in these complexes. The complexes are characterized by low stability. The formation of inner-sphere complexes is due to covalent bonds; there are no water molecules between the metal ion and the ligand. These complexes are highly stable. The stability of complexes depends on the nature of metal and ligand, as well as external factors (ionic strength, pH, temperature, etc.) (Senesi and Loffredo, 2005).

Humic Substances Perform Essential Ecosystem Functions (Orlov, 1990; Hayes and Clapp, 2001; Orlov and Sadovnikova, 2005):

1. The accumulation function denotes the accumulation of carbon, nitrogen, phosphorus, sulfur, and other elements, including trace elements, in the forms of organic compounds in soils, as well as the accumulation of chemical bond energy.
2. The transport function denotes the formation of geochemical fluxes of mineral and organic substances. Most inorganic and organic compounds migrate in the soil profile and landscape in the form of associations with HSs.
3. The regulatory function is a complex function implemented at different organization levels of the ecosystem: the participation in the development of soil structure and water-physical properties; the participation in ion-exchange reactions and acid–base and redox

equilibriums; the regulation of conditions of the mineral nutrition of plants; the regulation of the thermal regime of soils via the effect on the reflectance of soils; and the regulation of the soil differentiation of chemical composition.

4. The protection function denotes the capacity of HSs to bind chemical elements or compounds into low-mobile or hardly dissociated forms and thus protect living organisms from their toxic action.
5. The physiological function denotes the direct specific effect of HSs as stimulators of growth and development of plants.

Chapter 2

Study Objects

Objects of study were selected in accordance with the set problems. To study humic substances and their interactions with metals, two groups of soils were examined. One group included soils of different natural zones for characterizing the effect of zonal conditions on the formation of humus acids and the development of their properties. This group consisted of soils of the taiga (podzolic soil subtypes), steppe (chernozem), and semidesert (sierozem) zones. The second group included soils from the taiga zone for revealing the effect of different factors on the properties of humus and products of its interaction with metals. These were (a) natural factors, (b) technogenic factors and (c) analytical factors. Among the natural factors, the effects of some chemical properties of soils and soil-forming conditions were studied. The soils of technogenic landscapes were compared with their natural analogue. Among the analytical factors affecting the interaction of humic acids with metals, the content of metals and organic substances and the acid–base conditions were considered.

The studied soils of the taiga zone of Russia included soils from different regions: podzolic and soddy-podzolic soils of spruce and pine forests in Murmansk oblast developed under the natural conditions of forest biogeocenoses and in the technogenic impact zone of aerosol emissions from the Severonikel smelter; podzolic and peaty-podzolic forest soils of a reserved zone in Tver oblast (the Central Forest Biospheric Reserve). The studied chernozems included old-arable soils: ordinary chernozems of Voronezh oblast and southern chernozems of Odessa oblast. The sierozems under study were represented by sierozem from Syria (Mesopotamia). The upper humus-enriched horizons and litters of forest soils were predominantly studied. To characterize the migration capacity of humic substances, soil samples taken throughout the soil profile (podzolic soil from Tver oblast, the Central Forest Biospheric Reserve) were analyzed.

The conditions of soil formation that control the accumulation of organic substances by plants and their consequent transformation are strongly different for the studied soils (Kononova, 1951, 1963; Aleksandrova, 1980; Orlov, 1990). The soddy-podzolic soil was developed in the zone of coniferous–deciduous forests, where climatic conditions are characterized by the mean monthly temperatures of –4 to –8°C in January and 15–18°C in July, the mean annual precipitation of 600–700 mm, the rainfall factor of 0.6–1.0 (Ponomareva, 1964; Dobrovol'skii, 1999; Kovda, 1973), the period of biological activity of 60–90 days (Orlov et al., 1996), the high plant biomass (about 100 t/ha), and the tree falloff of 3–4 t/ha annually (Rodin and Bazilevich, 1965). The soil under study occurred under fallow conditions for about 20 years. Its morphological features (the agrohumus horizon of up to 40–50 cm in thickness) and chemical properties differ from those of the soils in natural landscapes because of the long-term agricultural use.

The properties of the studied chernozem (Kamennaya steppe, Voronezh oblast) are determined by its formation under conditions of continental climate and steppe herbaceous vegetation, where perennial grasses with developed root systems predominate; roots make up 70–80% of their biomass of 10–30 t/ha (Rodin and Bazilevich, 1965). The mean annual air temperature is 5–6°C, the mean annual precipitation is about 500 mm, and the average period of biological activity lasts for 170 days (Afanas'eva, 1966; Orlov et al., 1996).

The studied sierozem is a representative of soils of the arid semidesert zone in Mesopotamia, Syria, where the mean annual air temperature is 16–18°C; the annual precipitation is 100–200 mm, and the sum of active (>10 °C) daily temperatures reaches 3500–5800 degree-days (Glazovskaya, 1973; Kovda, 1973; Ilaiwi, 1983; Abd Al-Kareem et al., 1994; Dobrovol'skii, 1999). The biomass of desert-steppe plants is 4.0–4.5 t/ha (Rodin and Bazilevich, 1965).

Chapter 3

STUDY METHODS

The following methods were used in the work:

1. Methods for determining general chemical properties of studied soils;
2. Methods for the analysis of water extracts from soils;
3. Methods for the obtainment of humic acid preparations;
4. Methods for the examination of humic acid properties;
5. Methods for the study of the properties of water-soluble organic substances;
6. Methods for the study of the interactions of water-soluble organic substances with metals.

1. Methods for determining general chemical properties of studied soils. General properties of soils were determined according to the procedures described by Vorb'eva in the book *Chemical Analysis of Soils* (2006). The values of pH_{H2O} and pH_{KCl} were determined by ionometry; the content of organic carbon was determined by the Tyurin method (oxidation with potassium dichromate); potassium was determined by flame photometry; calcium, magnesium, copper, and zinc were determined by the atomic absorption method; the contents of gypsum and $CaCO_3$ were determined by the thermal method.

2. Methods for the preparation and analysis of water extracts from soils. Water extracts were prepared at a soil : water ratio of 1 : 5 for mineral horizons and 1 : 20 for organic horizons. In water extracts, total acidity was determined by titrimetry; pH was determined by ionometry; conductivity was measured by conductometry; anions SO_4^{-2}, NO_3^-, Cl^-, and PO_4^{-3} were determined by chemical methods (Vorob'eva, 2006); metals were determined by IPC on an Agilent 7500A spectrometer. Carbon of organic and inorganic compounds in water extracts was determined on a TOC-VCPH analyzer.

3. Methods for the obtainment of humic acid preparations. After the separation of organic residues (roots, leaf fragments, etc.), soil samples were dried at room temperature and triturated to <2 mm, and a sample of 1.0 kg was selected. In carbonate-containing soils (chernozem and sierozem), carbonates were degraded with a 10% HCl solution. The soil samples were decalcified with a 0.05 N H_2SO_4 solution (at a soil : solution ratio of 1 : 10) and, after standing overnight, filtered. Humic acids were extracted from the residue by successive double treatment with a 0.1 N NaOH solution at a soil : solution ratio of 1 : 10 (after the first treatment, standing for 6–8 h and filtration; after the second treatment, standing for 16–18 h and filtration). The first and second filtrate portions were combined.

Coagulation of dispersed mineral colloids: 250 ml of a saturated ammonium sulfate solution was added to 1 l of alkaline extract; the mixture was thoroughly stirred, left to stand for 1 h for the complete coagulation of colloids, and centrifuged on a flow-through centrifuge at 2500 rpm. Humic acids were precipitated in the purified alkaline extract by adding a 1 N H_2SO_4 solution to pH 1.1–1.2; the precipitate was separated and washed. The humic acid preparations were subjected to electrodialysis to purify from salts.

The preparations were dried in a vacuum drying oven.

4. Methods for the examination of humic acid (HA) properties. The content of ash in a HA sample was determined by incineration at 800°C. The elemental composition of HAs (C, N, S, and H) was determined on a Hewlett Packard 185 HB analyzer; oxygen was determined by the difference.

The molecular-weight distribution of HAs was determined on a column 9 mm × 60 cm in size (Pharmacia, Sweden) packed with Sephadex G 75 gel using a 0.03 M Tris–HCl buffer (pH 7.2) as eluent at an elution rate of 0.1 ml/min. The molecular weights of the fractions were calculated from the empirical Determan equation (Determan, 1968) based on the relationship between the HA molecular weight and the eluent outflow from the column.

IR spectra of HAs were recorded on an IKS-29 spectrophotometer using the KBr technique (Orlov and Grishina, 1981; Orlov and Osipova, 1988).

Recording of the ^1H NMR spectra of HAs: a sample of HA preparation of ~20 mg was dissolved in 0.5 ml of a 0.3 M NaOD solution in D2O after the addition of 4–5 drops of 40% NaOD and transferred in a 5-mm NMR sample tube. ^1H NMR spectra were recorded on a Bruker Avance 400 spectrometer (frequency 400 MHz, signal-recording time 3.5 s, relaxation time 0.3 s).

The hydrophilic–hydrophobic properties of HA preparations were determined by hydrophobic interaction chromatography. An HA solution (fraction 1 kDa, pH 7.0) was put into a column of size 10 cm × 9 mm packed with Octyl-Sepharose CL-4B gel, and the hydrophilic fraction was eluted with a 10% $(NH_4)_2SO_4$ solution in a 0.03 M Tris–HCl buffer solution (pH 7.2) and then with a 0.03 M Tris–HCl buffer (pH 7.2). A stepwise gradient of the concentration of detergent sodium docecyl sulfate (SDS) in 0.03 M Tris–HCl from 0 to 0.3% was used for the elution of the hydrophobic fraction. Residual HAs in the column were washed with a 0.05 M solution of Trilon B at an elution rate of 0.3 ml/min. The elution profile was obtained at 280 nm. The contents of HA components with hydrophilic and hydrophobic properties were thus determined.

Acid–base properties of HA preparations were determined by the back potentiometric titration of HA alkaline solutions (pH adjusted to 12.0, ionic strength 0.1) with a 0.1 M HCl solution on a Horiba F-16 pH-meter using a Horiba 6367-10D combined glass electrode. The titrant flow rate was 3 ml/h; the pH value was automatically recorded every 30 min. The dissolved carbon dioxide was removed by blowing CO_2-free air through the extracts during the titration. The temperature during the titration was maintained constant at 25°C.

The Gran function (Gran, 1952) with modifications (Takamatsu and Oshido, 1978) was used to determine the initial and end points of HA titration. To determine the points of equivalences of HAs, differential titration curves were plotted in coordinates V – ΔpH/ΔV. In this curves, the maximums were taken as HA points of equivalence. The averaged dissociation constants of acid functional groups in HAs were determined graphically for specific pH ranges from the relationship between pH and $log(a/1–a)$ and the Henderson–Hasselbalch equation:

$$\mathrm{pH} = pK_a + \log(a/1-a),$$

where a is the degree of dissociation of acid functional groups, and pK_a is their dissociation constant.

5. Methods for the study of the composition of water-soluble organic substances and metals bound to them. In water extracts from litters of forest soils, the fractionation of water-soluble organic substances (WSOSs) and metals bound to them (Cu, Ni) was performed using a set of independent methods. WSOSs were fractionated by molecular weight, and sorption and desorption capacity. The fractionation of WSOSs by molecular weight was performed using gel chromatography on Sephadex G-50 and G-25 analogously to the procedure described above.

The fractionation of WSOSs was performed by the Forsythe method. According it water-soluble substances were first sorbed on the activated carbon; then their different fractions were desorbed from the carbon by the sequential treatment with 0.1 N HC1, 90% water-acetone solution, and 0.5 N NaOH (Forsyth, 1947). Two first extracts include nonspecific organic substances: various organic acids, purine bases, and simples sugars are desorbed with 0.1 N HCl, carbohydrates and phenols are desorbed with water-acetone solution. Fulfvic acids are desorbed for carbon with 0.5 N NaOH.

The content and composition of individual low-molecular organic substances (with boiling points below 450°C) in water extracts from litter were determined using exclusion (gel permeation), partition, and ion-exchange chromatography and chromato-mass-spectrometry on a gas chromatograph with a mass-selective detector; the data were recorded and analyzed with an HP G 1034 C MS Chemstation. In the course of the analysis, the organic substances were extracted from the water extract with diethyl ether, reextracted with ethanol, and separated on a chromatographic capillary column. In the resulting solution, individual volatile organic compounds with boiling points below 450°C were identified from the mass spectra and quantified.

Nonvolatile low-molecular organic acids in the water extracts were determined by ion-exchange and ion-exclusion chromatography on a Kontron Model 421 instrument with an ultraviolet detector and on a Tsvet 3006 chromatograph with a conductometric detector.

The separation is based on the interaction between the solute components and the column resin. The substances to be analyzed are distributed between the stationary water phase within resin particles and the mobile eluent moving among packing particles. The ionized compounds are rapidly eluted from the column, and the un-ionized substances are retained and pass through the column more slowly.

The acid–base properties of water extracts from soils were determined by potentiometric titration: direct and back titration of soil extracts with a 0.01 N HCl solution and a 0.01 N NaOH solution, respectively, with the control of the pH level. The dissolved carbon dioxide was removed by blowing CO_2-free air through the extracts during the titration.

6. Methods for the study of the interactions of water-soluble organic substances with metals.

Preparation of water extracts from soils. Water extracts were prepared from triturated (<2 mm) air-dry soil samples at a soil : water ratio of 1 : 2 for mineral horizons and 1 : 5 for organic horizons. After shaking for 15 min and standing overnight, water suspensions were filtered through blue-ribbon filters.

In the extracts, carbon of dissolved organic substances was determined by oxidation with potassium dichromate; calcium, magnesium, copper, and zinc were determined by the AAS method; potassium was determined by flame photometry, and iron was determined by photometry.

Water-soluble substances with different molecular weights were separated by gel chromatography on Sephadex G-50 and G-25 and eluted with a phosphate buffer (pH 7.2). Solution samples for analysis were taken at the column outlet (Khmel'nitskii, 1981).

The content and composition of individual low-molecular organic substances in water extracts from litter were determined using exclusion (gel permeation), partition, and ion-exchange chromatography and chromato-mass-spectrometry on a gas chromatograph with a mass-selective detector; the data were recorded and analyzed with an HP G 1034 C MS Chemstation. In the course of the analysis, the organic substances were extracted from the water extract with diethyl ether, reextracted with ethanol, and separated on a chromatographic capillary column. In the resulting solution, individual volatile organic compounds with boiling points below 450°C were identified from the mass spectra and quantified.

Nonvolatile low-molecular organic acids in the water extracts were determined by ion-exchange and ion-exclusion chromatography on a Kontron Model 421 instrument with an ultraviolet detector and on a Tsvet 3006 chromatograph with a conductometric detector.

The separation is based on the interaction between the solute components and the column resin. The substances to be analyzed are distributed between the stationary water phase within resin particles and the mobile eluent moving among packing particles. The ionized compounds are rapidly eluted from the column, and the un-ionized substances are retained and pass through the column more slowly.

Potentiometric titration of model solutions and water extracts from the studied soils with solutions of metals.

Model solutions of organic substances of known composition (oxalic acid) and water extracts from the studied soils were titrated with solutions of metals ($Cu(NO_3)_2$, $Al(NO_3)_3$, $Fe_2(SO_4)_3$, $Zn(NO_3)_2$ $Pb(NO_3)_2$, $Hg(NO_3)_2$, $HgCl_2$).

In case of copper nitrate, potentiometric titration of model solutions and water extracts from soils was performed by a 0.013 M $Cu(NO_3)_2$ solution prepared with a 0.1 M $NaNO_3$ solution (to maintain a constant ionic strength). The pH measurements were accompanied by the measurements of the Cu^{2+} ion activity with a copper-selective electrode. The electrode was preliminary calibrated using standard copper salt solutions in the presence of 0.1 M $NaNO_3$ in the pCu range from 2 to 5. The calibration curve was linear with a slope of 29 in this range, which corresponded to the Nernst equation. The curve slope remained constant during the entire experiment.

A 1 M $NaNO_3$ solution (4 ml) was added to an aliquot portion (40 ml) of extract (or model solution) to create a constant ionic strength (0.1), and the extract was put on a magnetic stirrer. A 0.013 M $Cu(NO_3)_2$ solution was added to the extract by 0.05–0.1 ml portions. The extract pH was determined with an ESL 43-07 glass electrode 2 min later, and the pCu value was measured on an Expert-001 digital ionometer with an XC-Cu-001 copper-selective electrode.

The content of complexed copper was calculated from the difference between the total content of copper added (Cu_{tot}) and the content of free copper ions (Cu^{2+}) determined with the copper-selective electrode.

The titration of extracts with solutions of other metals—($Al(NO_3)_3$, $Fe_2(SO_4)_3$, $Zn(NO_3)_2$ $Pb(NO_3)_2$, $Hg(NO_3)_2$, $HgCl_2$)—was performed analogously to their titration with copper salt solutions. Only the pH levels were determined during the titration, and the $NaNO_3$ solution was not used.

Calculation of the parameters of complexation of copper ions with organic ligands (oxalate ions, water-soluble substances of water extracts).

The parameters of complexation of copper ions with organic ligands (oxalate ions, water-soluble substances of water extracts) were calculated: the complexing capacities and stability constants of the forming complexes were determined from the Langmuir equation, the Gran function (Gran, 1952), Henderson-Hasselbalch equation and Scatchard plot (Stevenson, 1994).

Determination of the acid–base buffer parameters of soil extracts. The acid–base buffer parameters of the extracts were determined by continuous potentiometric titration. For this purpose, a 0.01 N HCl solution was added by drops to an aliquot (40 ml) of the extract. The pH of the extract was determined using a glass electrode 2 min after the addition of each titrant portion. After a pH of 3.0 was established in the extract, it was titrated with a 0.01 N NaOH solution to pH 10.0. The dissolved carbon dioxide was removed by blowing CO_2-free air through the extracts.

Assessing the effect of copper ions on complexes of iron and aluminum ions with water-soluble organic substances. A copper solution was added to an extract aliquot; then, the content of free Fe(III) ions was determined by the sulfosalicylic method, and the content of Al(III) ions was determined with aluminon. The results were compared to the contents of free Fe(III) and Al(III) ions determined in extract aliquots by the same methods without addition of copper solution. Separate calibration scales with and without copper were created to take into consideration the possible effect of copper ions on the reactions between indicators and metals to be determined.

Free iron ions were determined according to the following procedure. A 20-ml extract aliquot was put in a measuring flask of 50 ml. When the effect of copper was studied, 0.84 ml of a 0.01 M $Cu(NO_3)_2$ solution was added with an electronic pipette; the mixture was stirred and left for 10 min. Then, 0.5 ml of a 10% sulfosalicylic acid was added and stirred; 15 ml of a 10% ammonia solution was then added; the solution was brought to the mark with distilled water and stirred again. The optical density of the solution was measured against a blank solution on a photocolorimeter at 440 nm after 15 min.

The determination of free aluminum ions was performed as follows. A 25-ml extract aliquot was put in a measuring flask of 50 ml. When the effect of copper was studied, 1.05 ml of a 0.01 M $Cu(NO_3)_2$ solution was added with an electronic pipette; the mixture was stirred and left for 10 min. Then, 1 ml of a 30% ammonium sulfate solution and 30 mg of ascorbic acid were added. The solution was thoroughly mixed and brought to the mark with an acetate buffer. The optical density of the solution was measured against a blank solution on a photocolorimeter at 540 nm after 30 min.

Chapter 4

General and Specific Properties of Humic Acids and Their Migration in a Soil Profile

4.1. Properties of Humic Acids from Soils of Different Natural Zones

Humic acids (HAs), isolated from upper humus-enriched horizons of three soils from different natural zones, served as the objects for this study. The soils are (1) abandoned old-arable clay loamy soddy-podzolic soil on moraine deposits (Leningrad oblast); (2) arable clay loamy chernozem on loesslike sediments of the Kamennaya steppe (Voronezh oblast); and (3) sandy clay meadow sierozem on loesslike sediments (Mesopotamia, Syria). General properties of soils were determined by standard procedures described in Chapter 3 (1). Preparations of HAs were isolated from the soils and analyzed according to Chapter 3 (3, 4).

Chemical properties of the soils studied. The highest contents of humus and exchangeable cations and the highest values of exchange capacity and base saturation were observed for the chernozem, which had a near-neutral reaction (Table 4.1). The long-term agricultural use of soddy-podzolic soil and the application of mineral fertilizers and manure resulted in the formation of an increased content of humus (3.9%); a slightly acid reaction (pH_{H2O} 6.1, pH_{KCl} 4.9); and high levels of cation exchange capacity, content of exchangeable bases, and base saturation (90%).

The meadow sierozem is characterized by a slightly alkaline reaction, high contents of gypsum and carbonates, the presence of which in the topsoil is related to the shallow occurrence of saline groundwater and the rise of salts with the capillary fringe to the surface. The content of humus in the soil is very low.

Table 4.1. Chemical properties of humus horizons (0-30 cm) of studied soils

Soil	C_{tot}, %	pH		H_H	Ca^{2+}	Mg^{2+}	V, %	Σ $CaCO_3$+ $MgCO_3$, %	Gypsum, %
		H_2O	KCl	mmol$_c$/100 g of soil					
Soddy-podzolic	2.3	6.1	4.9	2.4	16.0	6.8	90	-	-
Cherno-zem	5.4	7.2	6.1	-	68.8	16.6	100	nd	-
Sierozem	0.7	7.8	-	-	-	-	-	17	55.3

Note: (nd) not detected.

Properties of HAs from the soils studied. Elemental composition of HAs. The elemental composition of HAs reflects the humification and mineralize-tion of organic substances under specific biochemical conditions. The content of carbon in humic substances is an essential identification index. For three soils under study, this parameter value is close to 53 wt % of ash-free dry matter. Oxygen makes up the third part of HA weight. The content of nitrogen is 3.6–5.4% (Table 4.2).

Table 4.2. Elemental composition of HAs from studied soils (wt % of ash-free dry matter)

HA from soils	C	H	N	S	O	Ash, %
Soddy-podzolic soil	53.1	4.7	4.2	1.6	35.3	1.2
Chernozem	52.9	3.6	4.0	1.2	34.9	3.4
Sierozem	53.5	5.4	5.9	1.6	31.1	2.5

The found contents of carbon and nitrogen in the HAs from the soils studied are within the ranges reported for the HAs of different soils: 46–62% for carbon and 1–6% for nitrogen (Kononova, 1951, 1963; Orlov, 1974, 1990,

1993; Schnitzer, 1978, 1981; Ganzhara, 1986; Lobartini et al. 1992, 1997; Brady and Weil, 2002). They are related to the nature and structure of HAs and characterize them as a specific class of organic compounds differing from many individual natural organic substances, such as proteins, lignin, cellulose, etc. (Kononova, 1963; Orlov, 1974).

Parameters of HA elemental composition expressed in atomic percent are important characteristics of HAs (Table 4.3). When presented in this form, parameters of HA elemental composition reflect the actual substitutions of atoms in aromatic rings and lateral chains of HAs (Orlov, 1974, 1992). The elemental composition of the studied soils is similar to the average parameters of HAs from the soils of different natural zones calculated for large sets of samples (Stevenson, 1982; Orlov, 1990; Orlov et al., 1996).

The HAs of the studied chernozem differ from the HAs of two other soils in the higher carbon contents. Only in the chernozem HAs, the content of carbon exceeds that of hydrogen. These characteristic features of chernozem HAs were already noted in the literature (Kononova, 1944; Orlov, 1974).

The HAs from the arable soddy-podzolic soil have different contents of carbon (lower than that in the chernozem HAs by 10%) and hydrogen (higher by 16%).

These parameters are similar to the average composition of HAs in soddy-podzolic soils, although the content of total C_{org} in the studied arable soil is higher than that in the soddy-podzolic soils of Russian forest and meadow cenoses (Orlov, 1990, Orlov et al., 1996). This is caused by the long-term effect of the active agricultural use of the soil.

The sierozem has a low content of humus; its HAs are characterized by a relatively low content of carbon and a higher content of nitrogen compared to two other soils. This is a result of specific humification conditions in sierozems. Pedogenesis in the dry-steppe zone occurs under alternating periods of high biological activity in spring and its fading during the long hot dry period in summer.

Table 4.3. Elemental composition of HAs (atomic % ash-free dry matter)

HA from soils	C	H	S	N	O	Combustion heat, cal/g
Soddy-podzolic soil	38.0	40.2	0.4	2.6	18.9	4240
Chernozem	41.9	34.3	0.4	2.7	20.7	4170
Sierozem	36.3	44.1	0.4	3.5	15.8	4830

An active humification occurs in spring, but the simultaneous intense mineralization of organic substances results in the depletion of humus in sierozems and carbon in humus. The lack of water and high temperatures in summer restrict the activity of soil microorganisms and their rates of detachment and assimilation of nitrogen-containing components. The accumulation of nitrogen in sierozems is presumably largely ensured by nitrogen-containing heterocyclic protein derivatives (Martin and Haider, 1971; Schnitzer, 1978).

The proportions of C, O, H, and N in the HAs of the soils under study are of information value (Table 4.4). The C/N ratio is 15.4 and 14.9 in the HAs from chernozem and soddy-podzolic soil, respectively. These values are close to the average values: 11–14 for the HAs of podzolic soils and 15–19 for the HAs of ordinary chernozem (Kononova, 1951, Orlov et al., 1996; Bezuglova, 2001). The lowest C/N ratio (10.5) was found for the chernozem HAs. This agrees with the previously noted high mineralization rate in the semidesert zone, which is accompanied by the active detachment of C-containing fragments of peripheral chains in the HAs of these soils and the retention of N-containing components.

Table 4.4. Atomic ratios and degree of oxidation of HAs from studied soils

HA from soils	Atomic ratio				Degree of oxidation
	C/N	C/O	H/C (H/C_{cor})	O/C	
Soddy-podzolic soil	14.90	2.01	1.06 (1.72)	0.50	-0.06
Chernozem	15.42	2.02	0.82 (1.46)	0.49	+0.17
Sierozem	10.51	2.29	1.20 (1.79)	0.45	-0.34

As for the H/C ratio, it is lower than 1 only for the chernozem HAs. This fact is indicative of the developed aromatic component of the HA molecule and its benzoidicity related to the volume and composition of organic residues in the steppe zone and to the specific biochemical conditions of their decomposition.

The corrected ratios $(H/C)_{cor}$ supplement the concepts of HA structure. Their calculation allows the substitution of a significant part of hydrogen atoms in the HA molecule by oxygen atoms to be taken into account. With consideration for these substitutions, this parameter is calculated from the following equation (Orlov, 1992, 1993):

$$(H/C)_{cor} = (H/C)_{ini} + 2\ (O/C) * 0.67.$$

The H/C_{cor} ratio for the chernozem HAs is higher than 1, although it remains lower than the corresponding values for two other soils, which reflects the increased content of carbon in the chernozem HAs. This indicates the higher maturity of HA molecules in chernozem compared to those in soddy-podzolic soil and sierozem. The data agree with the idea that the H/C atomic ratio in HAs depends on the stage of humification: when the period of biological activity elongates, this ratio decreases because of the accumulation of more stable products (Orlov, 1990). The degree of oxidation is low for HAs from all studied soils.

The atomic ratios of elements in HAs are informative parameters of the structural organization of HA components (Table 4.5). From the $(H/C)_{cor}$ ratio, the proportion of carbon in aliphatic chains of HAs was calculated using the modified van Krevelen equation (Orlov, 1992):

$$C_{aliph} = \frac{(H/C)_{cor} - 2/3) \cdot 100}{4/3},$$

The predominance of aliphatic fragments (60–84% of total carbon) is a common feature of HAs from three soils studied. This points to developed lateral chains in HA molecules of three soils, especially in the chernozem HAs, in which the portion of aromatic structures exceeds their contents in the HAs of soddy-podzolic soils and especially chernozem by almost 2 times. The high degree of benzoidicity in the chernozem HAs reflects the ratio of organic matter humification and mineralization activities under the biogeochemical conditions of the steppe zone.

Table 4.5. Distribution of carbon between aliphatic and aromatic structures in HAs from the soils studied

HA from soils	C of aliphatic chains, %	C of aromatic structures, %
Soddy-podzolic soil	79	21
Chernozem	60	41
Sierozem	84	16

Molecular-weight distributions of HA particles in the soils studied. Tendencies in the molecular-weight distributions of HAs from the studied soils were revealed (Table 4.6). Three fractions with different molecular weight were separated from the HA preparations: the fraction of high molecular weight (≥74 000 Da), the fraction of medium molecular weight (4 000–74 000 Da), and the fraction of low molecular weight (<4 000 Da).

Table 4.6. Molecular-weight distribution of HAs from the soils studied

HA from soils	I fraction		II fraction (most frequent MW)		III fraction	
	MW (Da)	Proportion, %	MW (Da)	Proportion, %	MW (Da)	Proportion, %
Soddy-podzolic soil	≥74 000	18	4 000-74 000 (18 200)	73	<4 000	10
Chernozem	≥74 000	9	4 000-74 000 (23 500)	86	<4 000	5
Sierozem	≥74 000	27	4 000-74 000 (25 100)	69	<4 000	3

The medium fraction is predominant in HA preparations from all three soils. The most common contents of particles with medium molecular weights are relatively similar for HAs of three soil types (18 200–25 100). However, the degree of dispersion of HAs varies among the soils.

The molecular-weight distribution in the chernozem HAs is more homogeneous. Almost 86% of all humic substances are concentrated in the fraction 4 000–74 000 Da; the fractions with coarser and finer particles make up lower than 15% at the 2 : 1 ratio. These proportions could be related to the age of chernozems. It is supposed that the proportion of the largest molecules in HAs decreases and the portion of the molecules with intermediate size increases with increasing time of soil development (Ghosh and Schnitzer, 1980; Orlov et al., 1988; Orlov, 1990; Min'ko et al., 1991). The predominant formation of high-molecular-weight humic substances in chernozems is also favored by the period of water deficit, which facilitates the polycondensation of HAs (Kononova, 1963). The proportion of the low-molecular-weight fraction in the chernozem HAs is extremely low.

The observed tendencies agree with those revealed in the analysis of the elemental composition of the chernozem HAs. The ratio between the aromatic and aliphatic HA structures found from the parameters of HA elemental composition indicates a higher proportion of stable aromatic structures in the chernozem HAs compared to the HAs from two other soils.

The HAs of soddy-podzolic soil are more heterogeneous in terms of particles with different molecular weights compared to the chernozem HAs, as was also noted previously (Orlov, 1990). The fraction with medium molecular weight is predominant, but the content of the high-molecular-weight fraction (18%) is almost double that in the chernozem HAs because of the lower decomposition rate of some nonspecific compounds in organic residues. The content of the low-molecular-weight fraction in the HAs of these soils is also double that in the chernozem HAs. This can be related to the higher water content in soddy-podzolic soils during humification, when the removal of water as a by-product of HA condensation is restricted, which limits the growth of HA molecule (Kononova, 1963).

Even a more heterogeneous molecular-weight distribution is observed in sierozem HAs, where it has a bimodal character. The predominance of particles with medium molecular weights remains, and the high-molecular-weight fractions make up one fourth of HAs weight.

A heterogeneous composition of the high-molecular-weight fraction is observed in sierozem HAs. The active mineralization of HAs in these soils results in the decomposition of unstable organic components and a relative increase in the content of high-molecular-weight components, including lignins of different categories. In the nonhydrolyzable strongly bound component of sierozem HAs, they can be partially composed of substances of peptide and protein nature, because sierozem HAs differ from those of two other soils studied by a relatively higher content of nitrogen.

IR spectra of HAs from the soils studied. IR spectra provided information on the qualitative composition of HAs, the contents of their components, and atomic groups in HA molecules of three soils studied (Larina and Kasatochkin, 1972). The clearest differences were manifested between the spectral parameters of the HAs from chernozem and sierozem (Table 4.7).

The IR spectrum of the chernozem HAs contains a very broad and intense band at 3390 cm^{-1} characterizing the aliphatic component of HAs. Radicals of the peripheral part absorb in this region, and this band is due to the stretching vibrations of hydroxyl groups predominantly bound by hydrogen bonds. A weak absorption band at 2920 cm^{-1} due to the stretching vibrations of groups $-CH_2$ and $-CH_3$ in the aliphatic part of HA molecule is noted, as well as a weak absorption band at about 1460 cm^{-1} (1540, 1380 cm^{-1}) due to the deformation vibrations of groups $-CH$, $-CH_2$, and $-CH_3$. Stretching and deformation vibrations differ by the direction of the atom shift angle. The band in the region 1720–1700 cm^{-1} indicating the presence of carboxyl groups in the molecules of HAs confirms their classification among benzene polycarboxylic acids. The maximum intensity of this absorption band was typical for the chernozem HAs.

Table 4.7. Absorption bands of IR spectra of HAs from studied soils and their relative intensity

Frequency, cm^{-1}	Soddy-podzolic soil	Chernozem	Sierozem
3390	medium	medium	medium
2920	medium	weak	strong
1460	medium	weak	strong
1716	medium	strong	weak
1650	weak	weak	strong
1620	medium	strong	weak
1540	weak	weak	strong
1380	medium	medium	medium
1220	medium	strong	weak
1129	weak	weak	strong
1040	medium	strong	weak

High intensities were observed for the bands at 1620 and 1515 cm^{-1} due to the in-plane vibrations of C=C bonds in the aromatic fragment of HAs. This points to the highest degree of aromatization of the chernozem HAs compared to those of two other soils. A band of medium intensity at 1380 cm^{-1} is due to the symmetric deformation vibrations of the $\equiv C{-}CH_3$ groups and vibrations of the ionized carboxylic group, which indicates a partial substitution of hydrogen in the carboxyl groups of the chernozem HAs. The intensity of this band is similar in three HA preparations of the soils studied. The strong and broad band at 1220 cm^{-1} belongs to the vibrations of the C–O and partially –OH bonds in the carboxyl groups of carboxylic acids. A high intensity of this band is observed in the spectrum of the chernozem HAs compared to the HAs of two other soils. The weak band at 1129 cm^{-1} indicates vibrations of atoms in the C–OH groups of tertiary and secondary alcohols. The band at 1040 cm^{-1} is assigned to vibrations of the C–OH groups of primary alcohols. The high intensity of this band in the chernozem HAs indicates the high content of this group compared to the HAs of soddy-podzolic soil and sierozem.

The IR spectra of sierozem HAs. The presence of hydroxyl groups in the aliphatic fragment of sierozem HAs in the amount comparable to those in the HAs of chernozem and soddy-podzolic soil was confirmed by the presence of a broad and intense band at 3390 cm^{-1} in the IR spectrum. The content of groups –CH, $-CH_2$, and $-CH_3$ (2920 cm^{-1}) in sierozem HAs is higher compared to chernozem and soddy-podzolic soil, where organic substances can be more actively transformed by microorganisms. The content of carboxyl groups (1716 cm^{-1}) in sierozem HAs, on the contrary, is the lowest. An absorption band at 1620 cm–1 typical for the amide I (–CO–NH_2) group was noted in the IR spectrum of sierozem HAs, as well as an absorption band at 1540 cm–1 characteristic of the amide II group. This corresponds to the results of the elemental analysis of sierozem HAs, which confirmed their high nitrogen content.

The absorption bands at 1620 and 1515 cm^{-1} in the IR spectrum of sierozem HAs are of low intensity, which indicates their low degree of aromatization. These results correspond to the carbon ratio between the aliphatic and aromatic structures of HAs calculated from their elemental composition. Low-intensity bands at 1380 cm^{-1} are indicative of symmetric deformation vibrations of the ≡C–CH_3 groups in the HAs of both chernozems and soddy-podzolic soils. A medium-intensity absorption band at 1040 cm^{-1} is distinguished, which indicates vibrations in the C–OH groups of primary alcohols.

IR spectra of HAs from soddy-podzolic soil. A broad and intense band at 3390 cm^{-1} is distinguished in the IR spectrum of the HAs from soddy-podzolic soil, as well as in the chernozem HAs, which indicates the presence of hydroxyl groups in the aliphatic fragment of HAs. The slightly more intense absorption band at 2920 cm^{-1} is indicative of the higher content of groups –CH, –CH_2, and –CH_3 in the aliphatic fragment of the molecule. According to their IR spectra, the HAs from soddy-podzolic soil have the medium content of carboxylic groups.

By the degree of aromaticity, the HAs of soddy-podzolic soil (identified from the intensity of absorption bands at 1620 and 1515 cm^{-1} in the IR spectrum) occupy an intermediate position between the chernozem HAs and the sierozem HAs.

The ≡C–CH_3 groups and ionized carboxylic groups are identified from low-intensity bands at 1380 cm^{-1}, which can be assigned to the symmetric deformation vibrations in the ≡C–CH_3 groups, similarly to the HAs of two other soils. The medium-intensity absorption band at 1040 cm^{-1} is indicative of the C–OH groups of primary alcohols.

^{1}H NMR spectra of HAs from the studied soils. Additional information about the structure of HA molecules was provided by ^{1}H NMR spectroscopy, which allowed protons of different groups in HAs from the studied soils be identified. Chemical shifts in the ^{1}H NMR spectra were recorded against tetramethylsilane (TMS) insoluble in polar solvent, which served as an

external standard. The assignment of the ^{1}H NMR spectra of HAs from the studied soils was based on the studies of the last 30 years (Wilson, 1981; Yonebayashi and Hattori, 1989; Preston and Rusk, 1990; Frund et al., 1994; Stepanov et al., 1997; Conte and Piccolo, 1999; Kalabin et al., 2000; Chukov, 2001).

According to the positions of H atoms determined from the ^{1}H NMR spectra, four types of groups were distinguished in the structure of HA fragments (Table 4.8).

Table 4.8. Distribution of protons in different components of HAs from the soils studied

HA from soils	Protons of different groups, %				
	Aliphatic groups			Methoxyl groups and carbohydrates	Aromatic groups
	γ–CH_n		β–CH_n and α–CH_n	CH_n–O	C_{ar}–H
	0-0.6	0.6-1	1-2.8	2.8-4.0	6.0-9.0
Soddy-podzolic soil	6.0	19.0	24.0	32.0	20.0
Chernozem	5.0	14.0	25.0	40.0	16.0
Sierozem	9.0	18.0	22.0	30.0	21.0

Methoxyl groups and carbohydrates are the predominant proton sources for the HAs of the studied soils. Their highest content in the chernozem HAs is due to the more active detachment of polysaccharides during humification and the increase in the oxidation degree of aromatic rings and the portion of free functional groups. This is accompanied by the lower portion of protons in the aromatic rings of the chernozem HAs, which indicates a high substitution degree of aromatic structures in the chernozem HAs compared to the HAs from sierozem and soddy-podzolic soil. The total relative content of protons in aromatic rings, methoxyl groups, and carbohydrates of HAs is similar for three soils (about 50%). The other half of the protons occurs in different functional groups of different orientation with respect to the aromatic nucleus. Similar contents of protons in aliphatic groups $-CH_3$, $-CH_2$, and $-CH$ or $-COOH$ (which make up one fourth of the HA molecule) occur in the β and α positions with respect to the aromatic ring. According to the ^{1}H NMR spectra, a smaller amount of protons is in the $-CH_3$ groups in the γ position with respect to the aromatic ring (14–19%), and their lowest proportion is found in the chernozem HAs.

IR and ^{1}H NMR spectra complement each other in the characterization of the HA structure in the soils studied. The sum of aromatic groups with methoxyl groups and carbohydrates, rather than the aromatic groups alone, prevails in the HAs of the studied soils. In these terms, the chernozem HAs are superior to the HAs from two other soils. The contents of aliphatic groups of specific (α or β) orientation with respect to the aromatic nucleus are similar. The higher proportion of aliphatic fragments in the HA molecule of soddy-podzolic soil is attested by broad multipletes with σ 0.6–1.0 ppm in the ^{1}H NMR spectra and a relatively intense absorption band at 2920 cm^{-1} in the IR spectrum.

Hydrophylic–hydrophobic properties of HAs from the studied soils. The presence of polar and nonnpolar groups in the HAs of the soils studied determines the amphiphilic properties of HAs, i.e., the different abilities of these groups to interact with water. The distribution of amphiphilic fractions in the HAs of the soils studied and their hydrophobicity parameters (Hph) are given in Table 4.9.

The predominant formation of hydrophilic or hydrophobic components of humic substances is directly related to the humification of organic matter occurring *in situ* under aerobic or anaerobic conditions (Gel'tser, 1940; Aleksandrova et al., 1970; Lykov, 1973; Mirchink, 1976; Kravkov, 1978; Aleksandrova, 1980; Stepanov and Milanovskii, 1989; Samoilova et al., 1990; Milanovskii et al., 1993; Stepanov, 1995; Milanovskii, 2000, 2009).

Table 4.9. Amphiphilic properties of HAs (HA fractions > 1 kDa)

HA from soil	Fraction		Hph
	Hydrophilic fraction, %	Hydrophobic fraction, %	
Soddy-podzolic soil	33	66	2.0
Chernozem	32	68	2.1
Sierozem	36	64	1.8

The HAs of three soils under study manifested stronger hydrophobic properties. The carriers of hydrophobic properties are not polarized; these are strongly bound hydrocarbon fragments like alkanes (flexible saturated hydrocarobons); alkenes (linear ethylene hydrocarbons containing, e.g., a C=C bond); fatty acids; or aromatic polycyclic compounds (consisted of condensed benzene cycles), e.g., fragments of lignin or aromatic acids (Popov, 2004; Stepanov et al., 1997).

These components ensure the important role of HAs in the formation of soil aggregates and their water stability and restrict the migration of humic substances along the soil profile (Shein and Milanovskii, 2003; Milanovskii, 2009). The predominance of hydrophobic components in HAs reflects the general principle of humification: the selection and accumulation of microbiologically and thermodynamically stable products of organic matter transformation (Orlov, 1990; Ganzhara and Orlov, 1993).

By the degree of hydrophobicity, the HAs of the studied soils form the series: chernozem HAs > soddy-podzolic soil HAs > sierozem HAs. The highest content of hydrophobic fragments in the chernozem HAs found by hydrophobic interaction chromatography was confirmed by the data about the highest degree of benzoidicity in the HAs of this soil calculated from the elemental composition of HAs. Another proof for the high content of aromatic fragments in the chernozem HAs is provided by their IR spectra: a high intensity of absorption bands at 1620 and 1515 cm^{-1}, which are due to the in-plane vibrations of C=C bonds in the aromatic fragment of HAs. The content of hydrophobic fragments in the HAs of the studied soils is directly proportional to the content of carbon in HAs and inversely proportional to the H/C ratio.

However, the formation of organic components with hydrophobic properties in HAs of different soils is due to not only aromatic structures (benzene rings), but also uncharged and nonpolar aliphatic chains, on one hand, and aliphatic amphiphilic components, on the other hand. The content of components with hydrophobic properties in the HAs of the studied soils is higher than the content of components with hydrophilic properties. The content of aromatic structures, which mainly ensure the hydrophobic properties of HAs, is lower than the content of aliphatic components. The involvement of both hydrophobic and hydrophilic components in the formation of humus is a characteristic feature of soils of the autochthonous genesis. The chernozem HAs are characterized by a relatively higher degree of hydrophobicity (Hph 2.1); the proportion of aromatic structures (41%) is higher compared to the HAs of two other soils. The hydrophobic fraction of HAs from southern chernozem contains a significant amount of benzoic structures, as well as short and branched aliphatic chains. Carbohydrate fragments, carbonyl groups, and long aliphatic chains prevail in the hydrophilic fraction of HAs (see § 4.2).

The lowest degree of hydrophobicity is noted in sierozem HAs, which reflects the lowest content of hydrophobic fragments (aromatic structures), as well as a higher content of hydrophilic fragments (amino acids and carbohydrates) compared to the HAs of chernozem and soddy-podzolic soil. Thus, the hydrophobic properties of HAs are developed due to aliphatic components, whose contribution exceeds that of aromatic components by 3.2 times. Proteins, amino acids, and nucleic acids, whose higher content is indicative of an increased content of nitrogen in sierozem HAs, can affect the activity of mobilization of water molecules. There are also other proofs: an absorption band at 1620 cm^{-1} characteristic of the amide I ($–CONH_2$) group and an absorption band at 154 cm^{-1} characteristic of the amid II group were noted in the IR spectrum of sierozem HAs. The lowest degree of hydrophobicity of sierozem HAs is related to the high rate of humus mineralization, which creates specific proportions of aromatic and aliphatic components, as well as hydrophobic and hydrophilic components.

In the HAs of soddy-podzolic soil, the contribution of fragments of hydrophilic amino acids and other components to the development of hydrophobic properties and amphiphilic aliphatic groups (predominantly methoxyl, $–CH_3$, $–CH_2$, and –CH groups; alcohol and phenol hydroxyls; carboxyls; and carbonyls) is smaller than in the sierozem HAs, but it is also the determining factor (exceeds the effect of aromatic structures by 2.2 times). Partial loss in hydrophilic components of HAs from these soils can occur due to their removal from the upper horizon.

Thus, in different natural zones with the wide variation of the species composition of plant cover and the chemical composition of plant residues and the different activities of their microbiological transformation, a peculiar type of organic substances is formed: humic acids, whose elemental composition differs from that of any other organic substance. The chemical elements C, H, O, and N in HAs from soils of different natural zones enter into the composition of aromatic and aliphatic groups typical for the soils as specific natural formations. Stable hydrophobic humification products were predominantly accumulated in HAs of thee soils. They were found to have a number of properties in common: (a) polydisperse HAs with the predominance of particles 4000–74 000 Da in size (the most common particles were 18 000–25 000 Da in size); (b) the predominance of components with hydrophobic properties; (c) the presence of aromatic structures with double peptide bonds; (d) the predominance of aliphatic components; (e) the presence of aliphatic structures with different peripheral components (free and bound carboxyl groups, carbonyl groups, alcohol hydroxyls, and groups –CH, $–CH_2$, and –

CH_3); and (f) the leading importance of methoxyl groups and carbohydrates as sources of protons in HAs.

The specific ratios of atomic groups in HAs vary among the soils of different natural zones; they are determined by the humification conditions of their organic residues, the activity and character of assimilation of hydrophilic compounds (carbohydrates, sugars, etc.) and structural aliphatic fragments of macromolecules containing oxygen and nitrogen, the soil conditions (mineral components; texture; water content; and the acid–base, redox, and other conditions), and the landscape-geochemical conditions (topography, temperature and humidity conditions, biogeocenosis type, and others), which determine the possibility of fixing and removing water-soluble compounds.

Chernozem humic acids differ from those of soddy-podzolic soil and sierozem by the following properties: (a) higher content of atoms C and O in the HA molecule; (b) the normal molecular-weight distribution of HA particles and the highest content of the particles 4 000–74 000 Da; (c) relatively higher hydrophobicity; (d) higher content of aromatic structures; (e) relatively higher content of C in aliphatic chains; and (f) relatively higher content of carboxyl and methoxyl groups and carbohydrates.

Sierozem humic acids differ from those of chernozem and soddy-podzolic soil by the following properties: (a) relatively lower content of oxygen and higher contents of hydrogen and nitrogen in their elemental composition; (b) the highest content of aliphatic groups (to 84%); (c) higher content of high-molecular-weight particles (to one third); (d) the highest contents of amides I and II; and (e) the highest content of groups –CH, $–CH_2$, and $–CH_3$.

Humic acids of soddy-podzolic soil differ from those of chernozem and sierozem by the following properties: (a) the lowest content of nitrogen in the elemental composition; (b) the highest content of nitrogen in the elemental composition; and (c) the highest content of aliphatic groups with protons in the γ position with respect to the aromatic nucleus. Other parameters occupy an intermediate position between those of the chernozem HAs and the sierozem HAs.

4.2. Amphiphilic Properties of Humic Acids from a Chernozem

The following conditions for the separation of the humic substances were selected: a linear gradient of the detergent concentration from 0 to 0.3% SDS-na; the total eluate volume, 210 ml or about 30 volumes of the column (Amicon, L = 200 mm, D = 10 mm); the gradient was developed using an Ultrograd 11300 gradient mixer (LKB).

The experiments were conducted on humic acids isolated from the Ap horizon of a southern chernozem (Odessa oblast) by the procedure described in (Orlov and Grishina, 1981). We previously attempted to separate these humic acids by their degree of hydrophilicity using the salting-out method at different ammonium sulfate concentrations (Milanovskii et al., 1993). It was shown experimentally that the HA fractions that precipitated at different degrees of saturation with ammonium sulfate differ in their degree of hydration, limiting wetting angle on their hydrophobic and hydrophilic surfaces, solubility in water, and other properties. At low salt concentrations, HAs with more pronounced hydrophobic properties coagulate in solution. More hydrophilic HAs are precipitated at higher salt concentrations. The most hydrophilic HA fraction remains in solution at 100% saturated in ammonium sulfate.

The presence of amphiphilic components in the HAs studied was confirmed by their fractionation on Octyl-Sepharose CL-4B. The most hydrophilic HAs did not interact with the hydrophobic matrix; they migrated with the starting buffer flow in the void volume of the column to form fractions A, B, and C. An increase in the detergent concentration in the eluent (from 0 to 0.3% SDS-Na) resulted in the weakening of the hydrophobic interaction of the sorbed HA molecules with the matrix and their gradual elution from the column with fractions D, E, F, and G.

During the analysis, 10 mg of the initial HA preparation (HA_{total}) dissolved in 1 ml of 0.05 M Tris-HCl buffer with pH 8.2 was applied to the column. The elution was conducted at a flow rate of 20 ml/h for 2 h. The elution profiles were obtained by testing at 206 nm. After each fractionation, the gel was regenerated by the above procedure. Hydrophilic HAs (fractions A and B) and hydrophobic HAs (fractions E, F, and G) were selected and accumulated for further work by multiply repeating the experiment using a Bromma 2212 Helirac fraction collector (LKB).

The isolated HAs and the initial HA_{total} preparation were subjected to additional fractionation by gel filtration on Sephadex G-50 under the following conditions: a column of 1.84 x 74 cm (Pharmacia); buffer, 0.05 M Tris-HCl (pH 8.2) with 0.1% SDS; sensitivity, 0.2% T; elution rate, 10 ml/h; detection at 206 nm.

The molecular masses (MMs) of separate HA fractions calculated from the empirical equations are frequently reported in papers on gel filtration of

HAs (e.g., for Sephadex G-50: logM = 4.791 - 0.423V_e/V_0, where V_0 is the void volume, and V_e is the elution volume). However, such calculations (without account for the configuration and density of the molecules analyzed) give only an approximate idea of the true molecular mass values.

A more exact determination requires the equality (or similarity) of the densities and shape (Perrin) factors between the molecules of the HAs studied and the markers used for the calibration of the chromatographic column (Orlov and Milanovskii, 1987). Therefore, we relied on the characteristics of the amphiphilic HA fractions isolated by fractional precipitation (salting out), which were determined previously (Milanovskii, et al., 1993). Based on the results of measuring the specific viscosity of the HA solutions and the Simgh equation for (a) elongated and (b) flattened ellipsoids of revolution, we calculated the probable ratios between the axes of the HA particles (molecules, associates):

$$\text{(a) } h_{sp}/F = v = (L/D)^2/15(\ln^2 L/D - 3/2) + L/D^2/5(\ln^2 L/D - 1/2) + 14/15;$$
$$\text{(b) } h_{sp}/F = v = 16(L/D)/15\tan^{-1} L/D$$

where h_{sp} is the specific viscosity, F is the specific volume of the particles (calculated from the elemental analysis data), and L/D is the axis ratio.

The calculations showed that the particles of the amphiphilic HA fractions are similar in shape and have an axis ratio of about 2-2.5 under the specific analysis conditions. Therefore, the calibration curve for a chromatographic column with gel G-50 was plotted using the following markers: suinsulin *(L/D* = 2.05, MM = 5700 amu), cytochrome C (*L/D* = 1.90, MM = 12 000 amu), and chemotrypsinogen A *(L/D* = 2.45, MM = 25800 amu).

The chromatogram of the initial HA_{totaI} sample showed five peaks with molecular masses of about 22000, 21000, 12000, 9000, and 7000 amu. The hydro-philic HAs mainly include high-molecular compounds (for the system analyzed), and the hydrophobic HAs mainly include medium- and low-molecular components (Figure 4.1).

Using multiple fractionation with a fraction collector, we selected and accumulated hydrophilic HAs (the fraction with a molecular mass of about 21000 amu) and hydrophobic HAs (the fraction with a molecular mass of about 9000 amu).

To reveal the structural features of the HAs, the ^{1}H NMR spectra of the initial preparation (HA_{totaJ}) and amphiphilic fractions (Figure 4.2) were recorded on a Bruker WP-80 NMR spectrometer. The HA preparations were dissolved in a 0.1 N alkali solution prepared with heavy water.

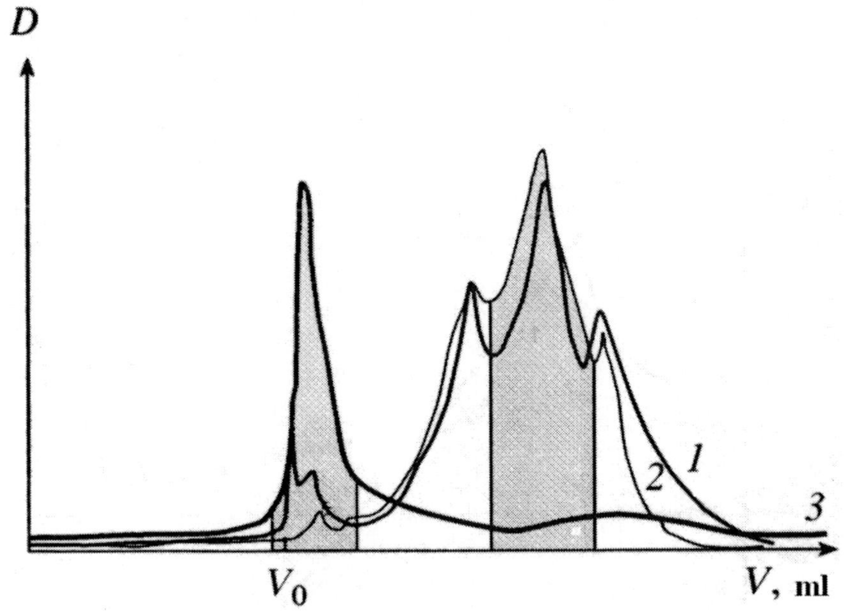

Figure 4.1. Gel chromatograms of humic acids: (1) HAtotal; (2) HAphob; (3)HAphil.

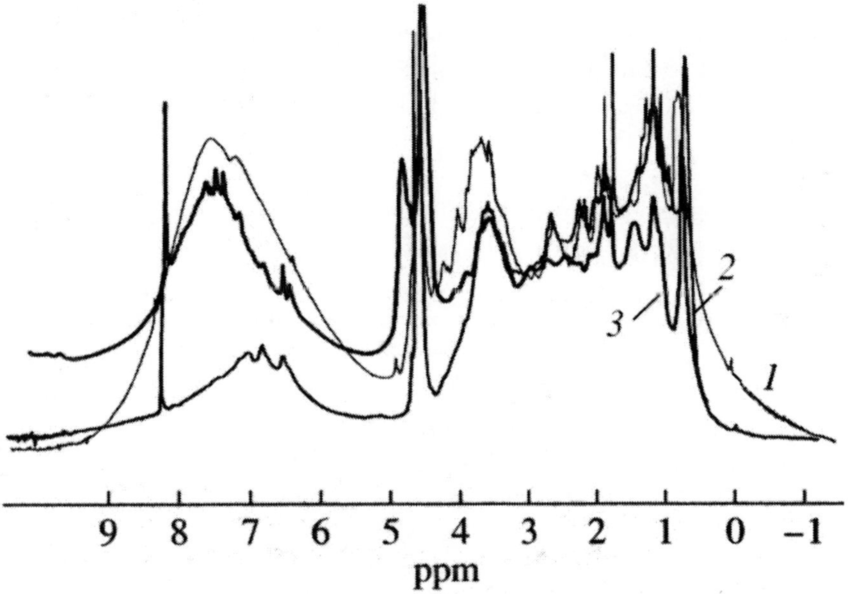

Figure 4.2. ^{1}H NMR spectra of humic acids: (J) HAtotai; (2) HAphob; (3) HAphil.

Table 4.10. Distribution of protons in humic acids

Sample	Portion of protons in HA fragments, %		
	Aliphatic chains	polysaccharides	benzoid rings
HA_{total}	66	11	23
HA_{phob}	61	9	30
HA_{phil}	71	18	11

The ^{1}H NMR spectrum of the initial preparation showed signals due to aliphatic protons in the region δ = 0.8-3.0 ppm (Table 4.10). The resonance signals at δ = 0.78 and 1.20-1.46 ppm are assigned to the methyl and methylene groups of alkyl chains in the HA molecules. A narrow doublet at δ = 1.84 and 1.94 ppm can be assigned to the protons of the CH_3-CO- groups, and a series of signals at δ = 1.94-2.60 ppm probably corresponds to the protons of the -CH_2-CO- groups. A multiplet at δ= 3.5-4.5 ppm is due to the protons of polyols (polyatomic alcohols). These are most probably the groups of carbohydrate fragments of HAs. A strong signal with a maximum at about 7.5 ppm from the protons of aromatic fragments is also observed in the range δ = 6.5-9.0 ppm. The presence of a narrow singlet at δ = 8.36 ppm in the high-frequency region is notable. Some authors assign this signal to the protons of the phenyl group, which is characterized by a local mobility with respect to other structural fragments of HAs. Other authors assign it to the format ion resulting from HA decomposition (Wilson, et al., 1983). The latter supposition is confirmed by the fact that the resonance intensity abruptly increases after the hydrolysis of HAs in 6 N HCl.

From the integrated signals in different spectral regions, the distribution of protons among the essential HA fragments (aliphatic chains, polysaccharides, and benzoid rings) was assessed (table). The proportion of aromatic protons in the hydrophobic humic acids (HA_{phob}) increased compared to that in the initial HA_{total} sample by almost one third, and the share of carbohydrate and aliphatic protons slightly decreased. In the HA_{phob} spectrum, a pronounced spin-spin splitting of the signal from the protons of aromatic compounds (probably, phenols) was observed at 6.7-7.7 ppm. The intensity of the band at δ = 4.89 ppm corresponding to the -CH=CH- groups also increased significantly. A series of proton signals typical for the CH_3-CH_2-CH_2-O- fragment (a triplet due to the CH_3 group at 0.61 ppm, a multiplet due to a CH_2 group at 1.74 ppm, and a triplet due to a CH_3 group at 2.99 ppm) was observed in the aliphatic spectral region of hydrophobic HAs. Two broadened multiplets at δ = 0.86 and 1.28 ppm were also observed in the same region, which could be assigned to the protons of the CH_3 and CH_2 groups. The similar intensities of these signals attest to the ramificated and short (C_2-C_4) structure of these radicals.

The share of aromatic protons in the hydrophilic fraction of humic acids (HA_{phil}) decreased by two times compared to the initial HA sample and by almost three times compared to the HA_{phob} fraction. This can be due to the lower content of benzoid structures in the HA_{phil} molecules. The intensities of the proton signals of the carbonyl groups and polysaccharides, on the contrary, increased by a factor of 1.5-2 compared to the HA_{phob}. The length of the aliphatic chains in the peripheral part of the HA_{phil} molecules increased up to 5-7 carbon atoms on the average.

Thus, the sequential fractionation of HAs from the southern chernozem allowed us to significantly increase the informational value of the proton spectra, to reveal the presence of separate groups and structural fragments with a high degree of probability, and to determine their content.

We earlier noted similar structural features of amphiphilic HA fractions when studying humic substances of meadow-gley soil (Stepanov et al., 1997). However, the obtained results give no information about the general structural features and formation rules of amphiphilic HA fractions in soils of different genesis and other natural materials. It is evident that the amphiphilic properties of HAs depend not only on the content of the hydrophobic or hydrophilic components but also, and to a significant degree, on the spatial configuration of HA molecules and their associates, the degree of ionization, the charge size, and the distribution density of charged groups on the surface of the particles.

4.3. Migration Ability of Humic Substances in Peat-Podzolic Soil

Investigations in recent years (Milanovskii, et al., 1993; Milanovskii, 2000, 2009; Stepanov, et al.,1997) has experimentally confirmed the presence of fractions of humic substances differing in their amphiphylity and solubility. This points to the possibility of migration of mobile humic substances in soil profiles. It is obvious that certain preconditions should exist for this: 1) the existence of a system of capillary pores and cracks favoring water infiltration; 2) blocking (at least for a short time) of sorption centers in the soil adsorption complex or a decrease in their activity (for example, during a period of intense rainfall); and 3) the particular composition, concentration, and pH of the soil solution; etc.

The presence of great amounts of humic and humi-clike compounds in natural waters, bottom sediments, yellow matter of seas, and other natural objects is an indirect argument in favor of possible migration of humic substances even in the presence of strong geochemical barriers. Many morphological features (pockets, cutans, tongues, soil cracks and fissures filled with organic material, humus traps, etc.) attest to the activity of humus migration in soil profiles. In this context, it is interesting to analyze the fractional composition of organic matter in such pedofeatures and in the surrounding soil mass.

For investigation of HS migration in the profile of peat-podzolic soil we analyzed HAs and FAs isolated from material filling fissures and from the genetic horizons of the studied soil. The choice of this object for our study was determined by the well-expressed polygonal fissuring of the soil, which is also typical of many other soils of eluvial-illuvial type. Vertically oriented fissures penetrated the soil profile to a depth of 70-90 cm and were filled with humus material of different degrees of decomposition and different colors (from light brown to black). The soil samples were from the material filling the fissures (from different depths) and from the soil horizons; two soil monoliths (35 x 35 x 50 cm) were cut from the walls of the trench.

At the first stage of laboratory study, the total contents of carbon (with the use of an AN-7529 analyzer) and nitrogen (according to the Kjeldahl procedure) were determined in the samples (Table 4.11).

In general, the concentration of organic matter clearly decreased down the soil profile, particularly upon the transition from the humus-eluvial horizon to the lower-lying mineral horizons. Nitrogen concentrations in plant fal-loff and litter horizons did not exceed 2%. The degree of humus enrichment in nitrogen was also low, and the C/N ratio exceeded 20 in the upper horizons. The values of this ratio decreased more than twice in the lower part of the soil profile, which was probably due to retardation of the destruction of nitrogen-containing organic compounds and an increase in the content of mineral nitrogen.

The organic matter content in the material of the fissure zone decreased down the soil profile much more slowly. Thus, the C_{org} concentration in the bulk soil mass decreased by almost five times in the illuvial horizon; in the material of the fissure zone, it decreased by 1.3-1.5 times at the same depth. The C/N ratio in the fissure zone ranged from 20 to 23 to a depth of more than 50 cm; the same values were determined in the upper humus horizon.

The appearance of the second maximum of organic carbon in the fissure zone at a depth of 20-25 cm should also be noted.

Table 4.11. Concentrations of carbon and nitrogen in the genetic horizons and in the material of fissure zones in the peat-podzolic soil

Horizon	Depth, cm	C_{org}	N_{org}	C/N
		%		
L	0-7	39.60	1.74	26.5
F		40.75	1.80	26.3
H		30.87	1.73	20.8
A1E	7-0(13)	2.41	0.13	21.6
E	10(13)-25	0.52	0.04	15.1
EB1	25-40	0.38	0.04	11.0
B1	40-65	0.21	0.03	8.1
Fissure	6-8	3.05	0.16	22.2
	8-10	2.79	0.14	23.2
	10-12	2.50	0.14	20.8
	12-14	2.20	0.13	19.7
	14-16	1.69	0.10	19.7
	16-18	1.70	0.10	19.8
	18-20	1.81	0.10	21.1
	20-22	1.96	0.11	20.7
	22-24	2.03	0.10	23.6
	24-26	1.93	0.11	20.4
	26-28	1.76	0.09	22.7
	28-30	1.50	0.08	21.8
	30-35	1.36	0.07	22.6

Table 4.11. (Continued)

Horizon	Depth, cm	C_{org}	N_{org}	C/N
		%		
	35-40	1.33	0.07	22.1
	40-45	1.10	0.06	21.3
	45-50	0.92	0.05	21.4
	50-55	0.84	0.05	19.5
	55-60	0.70	0.04	20.4
	60-65	0.50	0.03	19.4

These data allow us to assume that organic substances migrate from the upper horizons down the fissure zone, whereas frontal migration of humus in the soil profile is absent.

Then, preparations of HAs and FAs were extracted from the samples according to the procedure recommended by Orlov and Grishina (1981) and analyzed by different physicochemical methods: their elemental composition and concentrations of functional groups and carbohydrates were determined; acid hydrolysis of them was performed, and the concentrations and composition of amino acids in the hydrolyzate were determined; and the electronic absorption, infrared, combinative dispersion, and nuclear magnetic resonance (NMR, proton and C^{13}) spectra were obtained. The distribution of amphiphilic fractions and molecular-weight fractions of HAs from the fissure zone and from the soil genetic horizons were studied by hydrophobic chromatography and gel filtration.

Data on the elemental composition and major properties of HAs (Table 4.12) point to a significant difference between HAs extracted from the upper humus and lower mineral horizons of the soil. For example, HAs from the A1E and E horizons contain more carbon and acidic functional groups in comparison with HAs from the EB1 and B1 horizons; they are also characterized by positive degrees of oxidation, increased coefficients of extinction and chromaticity, and, hence, by a higher content of benzene fragments in the molecules (the C/H ratio is close to 1).

Table 4.12. Elemental composition and properties of humic substances isolated from the genetic horizons and the material of fissure zones in the peat-podzolic soil; fulvic acids were isolated according to Forsyth

Horizon/ object	Depth cm	Ash	C	H	N	O	H/C	O/C	ω*	α*	0.001% Humic Substances, $E_{465\ nm,\ 1\ cm}$	E_4/E_6	Functional groups, meq/ 100 g HS	
		%											COOH	OH
Humic acids														
A1E	7-12	1.1	52.16	4.22	3.00	40.54	0.96	0.58	+0.20	15.5	0.052	5.2	490	170
E	12-25	1.7	53.09	4.51	3.49	38.91	1.01	0.54	+0.07	10.8	0.050	5.2	470	130
EB1	25-10	2.0	49.27	5.06	3.20	42.44	1.22	0.50	-0.23	-	0.040	5.0	350	130
B1	40-65	2.4	47.93	6.21	3.36	42.50	1.53	0.66	-0.22	-	0.034	4.3	320	110
Fissure	10-20	1.3	52.22	4.43	2.79	40.57	1.01	0.58	+0.16	12.8	0.050	5.0	480	125
	20-40	2.2	53.25	4.56	3.35	38.84	1.02	0.54	+0.08	8.0	0.048	4.8	463	110
	40-70	1.9	52.33	4.39	3.07	40.20	1.00	0.50	+0.16	14.0	0.048	4.9	440	100
Fulvic acids														
A1E	7-12	2.40	45.20	4.53	4.18	46.09	1.19	0.77	+0.34	-	0.019	11.8	600	72
BTM	10-20	2.07	44.08	4.25	3.32	48.35	1.14	0.82	+0.48	-	0.016	11.3	610	76
Fissure	20-70	1.50	43.63	3.94	3.30	49.13	1.07	0.84	+0.60	-	0.016	11.0	620	74
Brook		1.10	43.84	3.72	3.20	49.24	1.00	0.84	+0.67	-	0.018	11.4	640	80

*ω is the degree of oxidation, and α is the degree of aromaticity (the portion of benzoic structures) of humic substances.

Humic acids from the lower part of the soil profile are close in composition and properties to prohumic substances and immature HAs extracted from plant residues (Milanovskii, 2000), which may be due to the specific conditions of in situ humification (oxygen deficiency, decreased biological activity, etc.). The elemental composition and properties of HAs isolated from the material of the fissure zone are similar to those of HAs from the upper soil horizons.

Fulvic acids isolated from the litter horizon and from water of a brook draining the studied territory are very close in their composition and properties, and this fact suggests the probability of migration of these substances in the soil and along the fissure zone.

Molecular weights of studied HAs and FAs were determined by the method of gel filtration in a G-50 sephadex (Pharmacia) under the following conditions: column of 1.84 x 76 cm (LKB); 0.05 M Tris-HCl (pH 8.2) with 0.1% SDS-Na buffer; sensitivity of 0.2% T; rate of elution of 10 ml/h; and detection at 206 nm.

As a part of investigated HAs there are fractions with molecular weight nearby 23000, 14000 and 3000 u. The ratios between these fractions in the genetic soil horizons change significantly down the profile: the content of the high-molecular components of HAs decreases with depth. In the lower mineral horizons, the fraction of HAs with a molecular weight of 10000-14000 u predominates.

The character of molecular-weight distribution of HAs in the fissure zone changes in a different way, as the portion of the high-molecular humus compounds increases downward the soil profile with a maximum at a depth of about 30-40 cm.

The number of amphiphilic fractions (division investigated HA spent on Octil-Sepharosa CL-4B) in the soil profile varies insignificantly: from 11-10 in the organic horizons to 10-9 in the mineral horizons. At the same time, the intensity of chromatographic maximums changes crucially. The indices of hydrophoby (hph) of HAs calculated as the ratio between the areas of hydrophobic and hydrophilic fractions can be used for the quantitative assessment of observed changes (Table 4.13). An assumption is made in this case about the closeness of extinction coefficients of individual fractions and, consequently, about the direct dependence of peak area on the amount of the substance.

The amount of hydrophilic substances in the humus from the soddy horizon of the soil is twice as high as the amount of hydrophobic components: the hph is 0.3.

Table 4.13. Degree of hydrophoby (hph) of humic acids from the genetic horizons and the material of fissure zones in the peat-podzolic soil

		Fractions		
Horizon	Depth, cm	hydro-philic	hydropho-bic	hph
A0	0-7	77	23	0.3
A1E	7-10(13)	54	46	0.9
E	10(13)-25	42	58	1.4
EB1	25-40	40	60	1.5
B1	40-65	30	70	2.3
Fissure	10-20	82	18	0.22
	20-30	83	17	0.20
	30-40	85	15	0.17
	40-50	82	18	0.22
	50-70	76	24	0.31

The ratio between these fractions in the transitional A1E horizon is close to 1, and hydrophobic fractions predominate in the lower part of the soil profile. For example, hydrophobic HAs comprise up to 70% of the total sum of fractions in the B1 horizon. These changes may be associated with worsening of aeration conditions down the soil profile, because the formation of hydrophobic humus is determined by the living activity of anaerobic microflora and proceeds inside soil aggregates (according to the hypothesis suggested by Shein and Milanovskii (2003). Hence, we can conclude that HAs from the soil horizons differ sharply in their composition and properties, and a decrease in the portion of hydrophilic fractions in their composition takes place in the deep horizons. This suggests the absence of (or very slow) frontal migration of HAs in the studied soil profile.

HAs isolated from the fissure zone are very homogeneous and close in composition and properties to HAs from the A1E horizon. This confirms the probability of humus migration down the fissure zone from the upper soil horizons in the form of colloidal solutions or in the form of coarser particles. The index of hydrophoby of HAs is low (0.2-0.3) along the entire studied fissure zone, which points to a rise in the portion of hydrophilic fractions in the composition of HAs down the fissure zone. The maximum accumulation and, simultaneously, the minimum total amount of amphiphilic fractions are observed at a depth of 30-40 cm.

The nature of amphiphility of HAs is of special interest. Traditional chemical and modern instrumental methods of analysis allow us to identify a wide range of structural components potentially responsible for the formation of hydrophobic and hydrophilic properties of humus compounds (Milanovskii, 2009; Popov, 2004; Stepanov, et al.,1997; Chukov, 2001; Cozzolino and Piccolo, 2002; Perminova et al., 2000).

To determine the structure of amphiphilic fractions of humus from the studied soil, we obtained ^{1}H-NMR spectra of conventionally hydrophobic and hydrophilic HAs from the A1E horizon . As noted above, eleven maximums are present on the chromatographic curve of HAs from the A1E horizon. The studied preparations included fractions I and II (hydrophilic HAs) and fractions VI to X (hydrophobic HAs). HAs accumulated with the help of a Bromma 2212 Xeriac (LKB) collector via multiple fractionation in Octil-Sepharosa CL-4B. The spectra were obtained on a Bruker WP80 NMR spectrometer (Germany). Preparations of HAs were dissolved in 0.1 N alkali solution prepared on the basis of heavy water. Integration of signals in different areas of the spectra made it possible to evaluate the proton distribution by the most important fragments of HAs: aliphatic chains, polysaccharides, and benzene rings (Table 4.14).

The part of aromatic protons in hydrophobic HAs increased by almost a factor of 2 in comparison with hydrophilic HAs, and the part of protons of carbohydrates and aliphatic fragments slightly decreased. Clearly expressed spin-spin splitting of proton signals of aromatic compounds (probably phenols) was observed in the spectrum of HA_{phob} under a chemical shift at 6.7-7.7 ppm.

Table 4.14. Distribution of protons in Yuma kazants from the A1E horizon

Sample	Portion of protons in components of humic acids, %		
	aliphatic chains	polysaccharides	Benzoic rings
HA_{phob}	44	37	19
HA_{phil}	45	45	10

Similar features in the structure of amphiphilic fractions of HAs were observed earlier for meadow gley and bleached brown soils of the Far East region, southern chernozem (Odessa oblast), and several mountainous soils of western Mongolia (Stepanov and Milanovskii, 1989; Stepanov, et al., 1997). These materials allow us to suppose that amphiphilic fractions of HAs from soils of different geneses and other natural objects have some common features. However, it is obvious that the amphiphility of HAs is determined not only by the amount of hydrophilic or hydrophobic components but also by the spatial configuration of HA molecules and their associates, the degree of ionization, the ionic charge, and the distribution density of charged groups on the particle surface.

Thus, we may conclude that humus compounds with predominant hydrophobic properties are formed in situ in genetic horizons of an iron-illuvial podzolic soil under conditions of a humid boreal climate. Micelles, aggregates, and clods should be the energetically favorable forms for the molecules of such compounds. This was confirmed by micromorphological observations: a significant part of humus compounds in the studied soils was weakly bound with mineral components; humus was localized in dark-gray humin and carbonic particles. Additionally, hydrophobic fractions of humus inherit the composition and structure of the initial plant material and are subjected less to the processes of oxidation and hydrolysis. This results in HAs enriched in nonpolar components (fragments of lignin molecules and aliphatic and polycyclic compounds), which is confirmed by the data on a elemental composition and the character of the proton spectra of hydrophobic HAs.

Hydrophilic fractions of humus are permanently formed during the process of humification of plant residues in litter subhorizons and in the upper humus horizons of soil. The frontal migration of mobile compounds down the soil profile is retarded under conditions of a slightly inclined soil surface, loamy soil texture, and a shallow depth of vadose water. The removal of organic matter may take place with lateral surface runoff (mostly during the snowmelt season upon water flows above the frozen soil layer according)or through the system of fissures and cracks. In the latter case, the migration of humic substances is limited by the length and diameter of fissures; the secondary maximums of humic substances may be formed at a depth of 20-40 cm.

Thus, the origin of humus in the material of fissures penetrating deep into the profile of peat-podzolic soil is related to the downward migration of organic com-pounds from the humus horizon in mechanical form and in the form of the most mobile fractions of humic and fulvic acids dissolved in the soil solution.

Chapter 5

Interaction of Copper Ions with Humic Acids from Soils of Different Natural Zones

Parameters of the complexing capacity of humic acids (HAs) for copper ions were experimentally determined for three soils (soddy-podzolic soil, chernozem, and sierozem) from different natural zones. The interaction of HAs from these soils with copper ions was studied with the methods described above (Chapter 3).

It was found that the addition of increasing amounts of copper salt to solutions of HAs from these soils with a corresponding increase in the Me : HAs ratio resulted in the successive formation of Cu–HA complexes with the participation of HA ligands of two types. Complexation parameters were calculated from the Langmuir plots drawn in accordance with the two-center model of complexation.

Parameters of complexation capacity with respect to copper ions were calculated for the ion-binding sites of the first and second types in HAs of the studied soils, as well as the corresponding stability constants of complexes (Table 5.1).

Q1 and K1 are parameters of complexing capacity for copper ions of ligands of the first type; Q2 and K2 are parameters of complexing capacity for copper ions of ligands of the second type in HAs of the studied soils.

The relatively low complexation capacities of copper ions with binding sites of the first type were similar in HAs of three studied soils: 24–26 $mmol_c$/100 g HAs (Table 5.1).

Table 5.1. Complexation capacities (mmol$_c$/100 g) of two ligand types in HAs of the studied soils with respect to copper ions and the stability constants of the complexes formed (pH 6.0)

Preparation	Complexation capacity of two types of HA ligands for copper ions			Logs of stability constants of HA complexes with Cu^{2+}	
	Q1	Q2	Σ	lgK1	lgK2
Soddy-podzolic soil HAs	25	66	91	5.3	2.9
Chernozem HAs	24	75	99	5.9	3.2
Sierozem HAs	26	59	85	4.8	2.8

The stability of complexes with ligands of the first type was almost double that of copper ion complexes with HA sites of the second type (lg K 5.9–4.8). These strongly bound compounds could be inner-sphere complexes; the increased ability of their formation at the interaction of metal ions with HAs was repeatedly noted (Bartashevsky et al., 1971; Vincler et al., 1971; Cheshire et al., 1977; Bloom and McBride, 1979; Schnitzer, 1981; Logan et al., 1997). The functional HA groups bound to benzene rings and capable of retaining copper ions by coordination bonds could also be involved. An aromatic carboxyl group and an adjacent phenol-hydroxyl group or two neighboring carboxyl groups can form stable salicyl-like and phenolic cyclic structures with metal ions (Tipping, 2005). The hydrate shell of metal ion is partially or completely destroyed, and the functional groups of the ligand occupy the coordination positions of the metal by substituting water molecules. The process follows the ligand-exchange mechanism (Piccolo and Stevenson, 1982). The metal occurs in the anionic moiety of polyatomic acids (Aleksandrova, 1980). It is supposed that the formation of bidentate chelates containing two functional HA groups is more stoichiometrically profitable at low degrees of saturation (Stevenson, 1994; Demin, 1994). Data are available that at pH 3–7, copper ions are bound to aliphatic substitutes such as amino, malonic, and citric acids in aromatic structures with the formation of five- and six-membered chelate rings (Gregor et al., 1989; Town and Powell, 1993).

The capacity of HAs from different soils to form stable complexes of the first types with metal ions is primarily determined by the structure of HAs (Chen et al., 1977; Senesi, 1990; Orlov, 1992). This capacity was most manifested for the chernozem HAs, which were characterized by an increased content of aromatic structures compared to the HAs of other soils (by two times higher than in the HAs of soddy-podzolic soil and sierozem).

Complexes of the second type involve less active reactive donor sites, which form less stable complexes (Zhorobekova et al., 1991; Stevenson, 1994). The covalent bond of copper with HAs, which is formed at low Cu^{2+} concentrations, changes for an ionic bond at high metal concentrations (Piccolo and Stevenson, 1982). The formed compounds can be outer-sphere complexes formed by hydrogen and electrostatic bonds, when the metal ion retains its hydrate shell during the interaction with the functional (carboxyl, carbonyl) groups of HAs (Gamble et al., 1976; McBride, 1978; Schnitzer, 1981). These are most probably 1:1 complexes. The stability of metal complexes with ligands of the second type is also affected by acidification occurring during the complexation of metal with the ligands of the first types.

The sierozem HAs were characterized by a relatively lower stability of complexes formed with copper ions. The composition of the functional groups involved in complexation is important. There is reason to suppose an increased content of nitrogen-containing functional groups (e.g., NH_2) in sierozem HAs. Copper ions form no stable complexes with them; they have a higher capacity (compared to zinc and cadmium ions) of forming stable complexes with oxygen-containing ligands than with nitrogen-containing ligands (Senesi and Loffredo, 2005).

Regularities of copper distribution in the complexes with HA fractions of different molecular weights were revealed. The products of interaction of copper ions with solutions of HAs from the studied soils were subjected to dialysis using dialysis tubes with 1- and 10-kDa mesh size membranes. The fraction >10 kDa was separated by molecular masses using gel filtration on Sephadex G-75. The ratios between the copper contents in the separated fractions (fractions I, II, and III are the high-, medium-, and low-molecular-weight fractions, respectively) are given in Table 5.2.

In the HA fraction <10 kDa, copper occurs in the free state, hydroxo complexes, and other low-molecular-weight complexes with mineral and organic ligands. In the interaction products of copper ions with HA solutions, the predominance of copper in the fraction >10 kDa was noted for the HAs of chernozem and soddy-podzolic soil: 79 and 59%, respectively. In the interaction products of copper ions with sierozem HAs, the content of copper remained in the free state and in low-molecular-weight complexes was almost double that in the complexes with HAs >10 kDa (63%).

The high correlation between the content of copper and the optical density of solutions containing HA fractions indicated a close link of copper ions with the corresponding HA fractions from all three soils.

Table 5.2. Copper distribution in products of dialysis and molecular-weight fractionation

HAs of soil	Cu in the fraction <10 kDa (free and complexed, % of Cu_{tot})	Cu in the fraction >10 kDa (free and complexed, % of Cu_{tot})	Cu distribution (mg/l) among the HA fractions >10 kDa, % of the content in the fraction >10 kDa		
			fraction I	fraction II	fraction III
Chernozem	21	79	5	88	7
Soddy-podzolic soil	41	59	11	81	7
Sierozem	63	37	21	65	13

Among the copper compounds in the HAs of the studied soils, copper complexes bound to HAs of medium molecular weights were predominant, especially in the complexes of chernozem and soddy-podzolic soil (81–88%); the value for sierozem was lower (65%).

This was primarily due to the predominance of the medium-molecular-weight fraction in HAs of all studied soils. According to the content of copper bound to the high-molecular-weight HA fractions, the studied soils formed the following series: sierozem > podzol > chernozem. A similar tendency was typical for the copper complexes bound to low-molecular-weight HA fractions.

The revealed copper distribution among the molecular-weight fractions of HAs from the studied soils can be related not only to the quantitative ratio of HA fractions, but also to the properties of these fractions. The participation of different HA fractions from three soils in complexation with copper ions also depends on the quality of these fractions, i.e., their activity in complexation. The Cu/C ratio in the fractions of different molecular weights serves as a quantitative measure for their participation in complexation with copper (Table 5.3).

The low-molecular-weight HA fractions from all three soils were most active in complexation, especially those from the sierozem HAs. They bound more copper into complexes by 4–5 times compared to the high- and medium-molecular-weight fractions from the same soil and by 2.5–3 times compared to the low-molecular-weight fractions from two other soils.

Table 5.3. The Cu/C ratios in HA fractions with different molecular weights

HA preparation	Cu/C ratio in HA fractions with different molecular weights		
	fraction I	fraction II	fraction III
Soddy-podzolic soil	0.07	0.11	0.13
Chernozem	0.05	0.10	0.16
Sierosem	0.08	0.10	0.40

Data are available (Stepanov, 1995; Debska et al., 2007) that the degree of dispersion of HAs is related to their hydrophobic and hydrophilic properties: the higher-molecular-weight HA fractions manifest more hydrophobic properties, and the lower-molecular-weight fractions manifest hydrophilic properties. The correlation observed between the degree of dispersion of HAs and their amphiphilic properties allows revealing the structural features of each HA fraction affecting its reactivity with respect to copper ions. The high reactivity of the low-molecular-weight fraction compared to the high- and medium-molecular-weight fractions is related to its high content of hydrophilic components and the predominance of acid carboxyl functional groups compared to the other fractions.

The sierozem HAs are characterized not only by the increased content of copper in the high-molecular-weight fraction, but also by the higher reactivity with respect to copper ions compared to the analogous fractions of HAs from chernozem and soddy-podzolic soil.

Copper Distribution in Complexes with HA Fractions of Different Amphiphilic Properties

Hydrophobic chromatography was performed of the HA fraction >10 kDa obtained after the dialysis of interaction products of copper salt solutions with solutions of HAs from the studied soils according to the procedure described above (Chapter 3).

The amphiphilic properties of humic substances significantly affect the mobility of metal compounds in the soil through the decrease in capacity of sorbing metals bound to the hydrophilic components of WSOSs and through the hydrophibization of the solid-phase surface, which increases the binding strength of heavy metals to the hydrophobic fraction of WSOSs (Karavanova et al., 2007). The distribution of the amphiphilic HA fractions from the studied soils and the concentrations of copper bound to them are given in Table 5.4.

Table 5.4 Relative contents of copper in the hydrophilic and hydrophobic fractions of HAs from the studied soils

HAs of soils	Cu bound to the hydrophilic HA fraction, %	Cu bound to the hydrophobic HA fraction, %
Chernozem	65	35
Soddy-podzolic soil	55	46
Sierozem	42	58

In the HAs of chernozem and soddy-podzolic soils, the portion of copper bound to the hydrophilic fraction (65 and 55%, respectively) was higher than that bound to the hydrophobic fraction. Another tendency was observed for the sierozem HAs: the portion of copper bound to the hydrophilic fraction was predominant (58% of the total copper content in the HA preparation with molecular weight >10 kDa). This was related to the increased capacity of the hydrophobic sierozem fraction for retaining metal ions. The Cu/C ratio in the hydrophilic and hydrophobic components serves as a measure for this capacity (Table 5.5).

Table 5.5. The Cu/C ratios in complex compounds of the hydrophilic and hydrophobic HA components

HA preparation	Cu/C in complexes of the hydrophilic fraction	Cu/C in complexes from the hydrophilic fraction
Chernozem	0.20	0.05
Soddy-podzolic soil	0.35	0.14
Sierozem	0.22	0.17

However, the reactivity of all hydrophilic HA fractions with respect to copper exceeded those of the hydrophobic fractions. This was related to the nature of the hydrophilic fractions, which contain carboxylic functional groups capable of complexing with metal ions. The relatively higher content of copper bound to the hydrophobic fraction of sierozem HAs can be due to the presence of nitrogen-containing functional groups in some hydrophobic aromatic compounds.

Chapter 6

WATER-SOLUBLE ORGANIC AND ORGANOMINERAL CU AND NI COMPOUNDS OF PODZOLS

Soils, lysimetric waters, and water extracts from Al-Fe-humus podzols of the Kola Peninsula were studied. The peninsula territory represents the northern limit ol forest biogeocenoses at the boundary of the northern-taiga and forest-tundra subzones. Lichen and shrub-green moss spruce and pine forests are predominant or the plain. The coniferous forests are usually thinned, with relatively stunted (12-16-m) trees. The phytomass reserves of these biogeocenoses are no more than 100 t/ha. The soils in the region under study include Al-Fe-humus podzols developed on glacial morainic sandy deposits containing up to 80% boulders. The soil profile composition is A0-A2-Bhfa-BC-C. The soils studied are characterized by deep (7-10-cm) forest litter, almost no humus horizon, and a well developed iron humus-illuvial horizon. We analyzed spruce and pine forest soils located 80-100 km from the Severonikel smelter that are slightly subjected to its aerosol wastes. These samples characterize the soils of conventional background. Soil samples were taken at 7 km from the smelter and characterize the soils of technogenic areas.

The soil samples to be analyzed were taken from horizons. The litter samples were not subdivided into subhorizons. The water extracts were obtained at a solid phase-to-water ratio of 1 : 20 for the litter and 1 : 10 for the mineral soil horizons. Ca, Mg, K, Cu, and Ni were determined by ICP-AAS; ions HCO_3^-, Cl^-, and SO_4^{2-} were determined by titrimetry. The ionic composition, pH, and total carbon were determined in water extracts. The total Cu and Ni and their mobile compounds in the H_2O and 1 N CH_3COONH_4 (pH 4.8) extracts were determined.

Water-soluble organic substances (WSOSs) were fractionated using a combination of methods, which are frequently used for the analysis of organic and organomineral substances in soils and natural waters. They were selected, because organic substances are the major components of the liquid phase in the soils, and their interaction with metals determines the material composition of their compounds. Metal compounds were fractionated by their ability to be sorbed on activated carbon, by molecular weight. These methods are independent, and each of them is not strongly selective and has some specific errors. However, when used together, they reveal common features and versatile characteristics of the substances studied. Water-soluble substances were fractionated in four replicates.

The fractionation water-soluble organic and organomineral compounds was fulfilled according Forsyth method.Water-soluble substances with different molecular weights were separated by gel chromatography on Sephadex G-50 and G-25 and eluted with a phosphate buffer (pH 7.2). Solution samples for analysis were taken at the column outlet (Khmel'nitskii, 1981).

The content and composition of individual low-molecular organic substances in water extracts from litter were determined (chapter 3).

The Properties of Organic Substances of Water Extracts

The water extracts from the litter and other genetic horizons of the studied Al-Fe-humus podzols contain low concentrations of all chemical elements (Table 6.1), which characterizes them as ultrasweet (Alekin, 1948). This is typical for soils of the extreme northern regions and is related to the low degree of decomposition of plant residues, the acid reaction of the soils, and the removal of the soluble mobile products of weathering and pedogenesis. Kovda (1973) classified these waters as ultrasweet iron-fulvate waters.

The waters from the test soils have an acid reaction; they differ in pH between the spruce and pine forests only slightly: 3.9-4.6 and 4.2-4.8, respectively. The proton activity in the waters of the spruce forests can exceed that of the pine forests by almost an order of magnitude.

The parcellar composition of the biogeocenosis frequently affects the acid-base properties of the soils to a larger extent than their belonging to different biogeocenoses. For example, the moss and cowberry parcels of the same biogeocenoses differ significantly (Lukina and Nikonov, 1996).

Table 6.1. Composition of water extracts from Al-Fe-humus podzols ($\times 10^{-3}$, mmol$_c$/L; C_{org}, mg/l)

Horizon	Depth, cm	pH_{H_2O}	C_{org}	Ca^{2+}	Mg^{2+}	K^+	HCO_3^-	SO_4^{2-}	Cl^-	Cu^{2+}	Ni^{2+}
Spruce forest											
A0	0-7	4.64	308	300.0	477	117.8	3.1	124.8	5.3	0.61	0.21
E	7-15	4.28	104	98.4	16.8	33.3	nd	120.0	nd	0.10	0.09
Bhfa	15-23	5.17	58	91.5	19.5	51.3	"	118.2	"	0.15	0.24
BC	23-38	5.50	nd*	109.3	17.2	32.5	"	125.0	"	0.12	0.13
C	38	5.53	"	92.1	14.9	20.5	"	121.6	"	0.13	0.32
Pine forest											
A0	0-2	4.79	186	290.4	248	71.8	2.2	185.7	8.7	0.53	0.28
E	2-5	4.63	44	113.0	27.2	61.5	nd	137.0	nd	0.26	0.15
Bhfa	5-20	5.24	19	96.1	31.0	51.3	"	116.4	"	0.23	0.32
BC	20-42	5.60	nd	63.2	23.1	43.6	"	129.1	"	0.15	0.09
C	42	5.57	"	53.6	19.4	32.9	"	124.0	"	0.19	0.07

* nd—not determined.

In the water extracts, the sum of cations exceeds the sum of inorganic anions (Table 6.1). The electrical neutrality of the solutions is ensured by organic acid anions. The total content of WSOSs in the litters of the spruce and pine forests varies significantly (100-308 and 60-186 mg/l, respectively). The content of carbon in the waters from the podzolic horizon decreases compared to the litter by 2-4 times.

The differences in the WSOS content between the litters of the spruce and pine forests are related to the biological cycle of carbon, which includes the volume, structure, and productivity of the phytomass; the falloff composition; the amount, diversity, and activity of the microorganisms; and the composition of the root exudates and the microbial, plant, and animal metabolites.

The reserves and the annual production of the phytomass are similar in the spruce and pine forests, but the amount of falloff is lower in the pine forests (Ushakova, 1999). The chemical compositions of the phytomass and mortmass differ between the spruce and pine forests, because the spruce forests more intensively extract Si, Ca, Mg, Mn, K, and N from the soil; Al and Na are more intensively extracted by the pine forests. The consumption of elements for the spruce growth is almost threefold greater than for the pine growth (130 and 68 kg/ha, respectively). The rates of the falloff decomposition differ only slightly. However, the spruce roots decay more slowly than the pine roots. According to Ushakova (1999), pine forest litters have lower values of total carbon, a lower C_{ha}: C_{fa} ratio (0.6 compared to 0.8), and a lower portion of nonhydrolyzable residue. These factors favor the development of a less acid reaction in the podzols of pine forests compared to spruce forests. The less acid reaction and the higher calcium content in the pine falloff can create more favorable conditions for microorganisms, which enhances their ability to decompose organic residues.

From the data of Kaurichev et al (1996), the total WOS carbon in the lysimetric waters of northern taiga can run to 200-260 mmol/l. The authors attribute the high content of organic substances in the soil waters and the relatively large loss in weight of the litter per season to the intensive mobilization of organic substances in the northern taiga.

Rode (1941) explained the high content of WSOSs in the falloff and litters of coniferous forests (134-701 mg/100 g litter) by the fact that the conifer-moss falloff supplies the largest amount of organic substances into solution at the early stage of decomposition. They can include not only compounds readily released during the decomposition of organic residues but also compounds washed down from the surface of leaves and needles (Mina, 1967; Lukina and Nikonov, 1996; Lysimeters in Soil Studies, 1998; Lukina et al., 2002). Water from the undercanopy usually contains organic exudates from leaves and needles and microbial metabolites. This is apparently typical not only for coniferous forests. For example, water from the undercanopy in oak forests of the Middle Sikhote-Alin contains 10-40 mg/l C_{org}, while its content in the atmospheric precipitation on open areas is no more than 1 mg/l (Arzhanova et al., 1981; Elpat'evskii and Lutsenko, 1990). Rainwater on open areas of spruce and pine biogeocenoses of the Kola Peninsula contains about 2-4 mg/l organic carbon, its content in the water from trees is 7- to 25-fold higher (Lukina and Nikonov, 1996). Thus, the load of organic substances in the water from the undercanopy that falls on the soil has a similar effect to the water passed through the litter horizon under these conditions.

The organic substances in the water extracts from litters of the Al-Fe-humus podzols are heterogeneous with respect to their molecular weight. Two predominant groups were isolated: high-molecular substances (MW > 19 000 Da) and low-molecular ones (MW < 2200 Da). No appreciable amounts of substances with MW 2200-19000 were manifested in the molecular-weight distribution curve. The total amount of carbon in the organic substances of the fractions separated was taken as 100% (Table 6.2).

Table 6.2. Molecular-weight distribution of the organic substances in the water extracts from the litters of the Al-Fe-humus podzols

Sephadex	Molecular weight	% of C_{total} in the water extracts
	Spruce forest	
G-50	>20000	24
	19300-19000	10
	2200-1700	60
	1700-1200	6
G-15	1010	-
	770-680	-
	Pine forest	
G-50	>20000	25
	19300-19000	6
	2200-1700	61
	1700-1200	8
G-15	1010	-
	770-680	-

Its value falls within the permissible range of organic substances in the water extracts from the soils studied before fractionation. The former WSOS group was subdivided into subgroups with MW > 20 000 and 19 000-19 300 Da; the latter group was subdivide into groups with MW 2200-1700, 1700-1200, and < 1200 Da. The share of low-molecular substances in the WSOSs is double that of high-molecular substances in both the spruce and pine forests (66-69 and 31-34%, respectively). They are classified as nonspecific organic substances.

The spruce and pine forests do not differ in the content of these substances in their water extracts. High-molecular compounds make up about one third of the total carbon and are classified as nonspecific compounds of fulvic-acid nature.

Additional information about specific and nonspecific organic compounds in the water extracts from the soils studied was derived using Forsyth's method involving the sorption of water-soluble organic substances on activated carbon followed by the desorption of separated fractions (Table 6.3).

Table 6.3. Specific and nonspecific organic substances (by Forsyth) in the water extracts from the litters of the Al-Fe-humus podzols ($C_{fraction}$, % of C_{total} in the water extract)

Sample	C, mg/l of the water extract	Fractions		
		nonspecific organic substances		fulvic acids
		0.1 N HCl	acetone-water	0.5 N NaOH
Spruce forest	308	9.8	59.8	28.0
Pine forest	186	8.0	67.7	22.6

The nonspecific and specific organic substances were reextracted by three consecutive treatments. The 0.1 N HCl solution presumably desorbs amino acids, purine bases, and some carbohydrates. Phenols and most carbohydrates pass into a 90% water-acetone solution. Subsequent water extracts remove polysaccharides. Fulvic acids are desorbed with a 0.1-0.5 N HCl solution.

The results of Forsyth fractionation also point to the predominance of nonspecific low-molecular substances in the water extracts from the litters of the soils studied. They make up 72-77% of the WSOSs in both the spruce and pine biogeocenoses. The major fraction (60-68% of the WSOSs) is desorbed by the water-acetone solution; it presumably consists of carbohydrate and phenolic components. Hydrocarbons and phenols are essential constituents of the litter liquid phase in taiga soils. For example, phenolic compounds constitute up to 45% in the water-acetone extract from the activated carbon on which the organic substances from the water extracts of the northern-taiga soils were sorbed by the Forsyth method (Kaurichev et al., 1991; Lysimeters in Soil Studies, 1998). Phenols, polyphenols, and tannides are relatively stable compounds with MW 500-3000 and a large number of phenol groups.

It is supposed that phenolic compounds are largely inherited by litters from leaves, needles, and falloff (Shido and Kuwatsuka, 1977; Wang et al., 1967). Lukina and Nikonov (Lukina and Nikonov, 1996) found polyphenols, as well as low-molecular carboxylic acids and carbohydrates, in the under-canopy spruce and pine forests of the Kola Peninsula. Their formation from the hydrolysis and microbial decomposition of the organic compounds prevalent in needles, primarily cellulose and other carbohydrates (which constitute up to 60% in needles), as well as of tannins (15%) and lignins (6%), is also supposed (Kretovich, 1971; Orlov, 1990).

No more exact data on the content of carbohydrates are available. It is known that carbohydrates are present in all soils, where they arrive with plant residues at a rate of 2-14 t/ha (Panikov et al., 1984). Although they are capable of relatively rapid transformation (assimilation by microorganisms, mineralization, or inclusion into the structures of newly formed organic substances), they partly remain hi soil waters.

The amount of organic substances extracted by a 0.1 N HCl solution is 6- to 8-fold smaller than that extracted by a water-acetone solution. The fulvic acids determined by Forsyth's method make up almost one third of the water-soluble organic substances (23 and 28% for the spruce and pine forests, respectively).

The absolute content of all the fractions separated in the spruce forests is almost double that in the pine forests. This fact is related to the differences in the content of total carbon in the water extracts from the litters, which depends on the differences in the falloff composition and the conditions of its decomposition, which were mentioned above. However, the relative contents of the separated fractions in both biogeocenoses are almost similar.

Thus, the molecular-weight distribution of the WSOSs and their ability to sorb on activated carbon attest to the predominance of low-molecular compounds in the litter WSOSs of the AI-Fe-humus podzols, as well as to the transformation of these compounds resulting in the formation of fulvic acids in the liquid phase of the soils studied. The amount of fulvic acids runs to 30% of the total WSOSs.

The predominance of low-molecular WSOSs in litters of forest soils was noted repeatedly (Elpat'evskii and Lutsenko, 1990; Zaitseva et al., 1987; Kaurichev et al., 1963, 1991; Kretovich, 1971; Lysimeters in Soil Studies, 1998; Christe, 2000; Stevenson, 1982; Tani et al., 1993, 1999, 2001). It was shown using sorption lysimeters that individual nonspecific compounds make up 57-80% of the total WSOSs in soddy-podzolic, bog-podzolic, and bog soils. Among them, substances with M W 400-410, 580-590, and 1500 Da were found, as well as smaller amounts of substances with MW 5500-5600 Da. The substances with MW < 1000 Da are predominant in the WSOSs resulting from the decomposition of plant residues (their fraction, by carbon, is 58%); they include fractions with MW of 260, 410, and 700 Da (Yashin, 1993). In the lysimetric waters collected under the humus-illuvial horizons of burozems in the Middle Sikhote-Alin, the organic substances contain 38-49% compounds with MW 500-800 Da and 18-29% compounds with MW 1000-5000 Da (Alekin, 1948; Elpat'evskii and Lutsenko, 1990). The authors believe that low-molecular substances include not only nonspecific substances but also fulvic acid monomers. Varshal el al. (1975) supposed that their share in natural waters is inversely proportional to the pH and reaches a maximum value at pH close to 4.

The organic substances in the water extracts from forest litters (including decomposed ones), fresh and composted plant remains (oak leaves, spruce needles, bark, green mosses, and epiphytic lichens from spruce trunks and branches), and the vegetative mass of 2-week-old barley seedlings labeled with ^{14}C were studied (Kaurichev et al., 1963, 1991; Lysimeters in Soil Studies, 1998; Yashin, 1993). Nonspecific compounds were predominant in all of the water solutions obtained; they made up 57-89% of the total carbon in the solutions.

We revealed the composition heterogeneity of the water-soluble organic substances and the effect of the biogeocenosis type, the geographical position of the soils, the time of the litter sampling and other factors. For example, fulvic acids were predominant in the WOSs from the peat-humus forest litter of the middle taiga in the fall; in the spring, their content became equal to that of the individual organic substances.

Oxalic, lactic, propionic, and succinic acids were found in the water extracts from the litters of the Al-Fe-humus podzols by ion-exclusion chromatography (Table 6.4). Oxalic acid was predominant. Its absolute content in the pine forests was 2-fold higher, and its relative content was 3-fold higher than that in the spruce forests. However, the total acid content was low; it did not exceed 0.1% of the total carbon in the water extracts.

These very simple carboxylic acids are mainly formed in plant tissues during respiration and fermentation processes (Kretovich, 1971). They are insignificantly transformed by biota and pass into water extracts from the soil studied. In solutions, they occur as salts and in the free state. The portion of free acids can be high because of the unsaturation of the sorption complex with alkali-earth bases.

The first evidence of the presence of nonvolatile low-molecular organic acids in extracts from litters and the decomposition products of forest falloff in northern-taiga soils were derived by potentiometric titration (Zaitseva et al., 1987; Rode, 1941; Skrynnikova, 1959; Khan, 1951) and partition chromatography (Aleksandrova, 1960; Kaurichev et al., 1963). In 1941, Rode calculated the degree of dissociation of nonvolatile acids from the potentiometric data for water extracts from forest soil litters and suggested the presence of salicylic, oxalic, and tartaric acid acids in these extracts (Rode, 1941).

Table 6.4. Low-molecular organic acids in the water extracts from the litters of the Al-Fe-humus podzols

Acid	Spruce forest		Pine forest	
	mg/l	% of C_{total} in the water extract	mg/l	% of C_{total} in the water extract
Oxalic acid	0.019	0.006	0.036	0.019
Lactic acid	0.014	0.005	0.011	0.066
Propionic acid	0.014	0.005	Absent	

Formic, oxalic, citric, lactic, fumaric, glycolic, acetic, benzoic, maleic, tartaric, malic, and succinic acids were identified in extracts from litters and plant material (birch and aspen leaves, spruce and pine needles, sphagnum moss, and matgrass) using paper, silica gel, and carbon chromatography. The highest content of acids was found in coniferous-forest litters (Candler et al., 1989; Yin et al., 1996).

Tani (Tani et al., 1993, 1999, 2001) studied water extracts from some forest soils (Andosols, podzols, cinnamonic forest soils, and red yellow soils) by HPLC and identified formic, lactic, oxalic, maleic, and citric acids. Formic and oxalic acids were predominant. Acetic acid was identified only in litter horizons. The total content of low-molecular organic acids was 4.6-37.2 mmol/kg.

The chromato-mass-spectrometric analysis of the water extracts from the litters of podzols of the Kola Peninsula showed the presence of individual volatile substances and a high portion of low-molecular organic acids.

The ultralow-molecular fraction (<500 Da), which made up about 0.1 % of the total WOSs, was analyzed. More than 70 individual organic compounds were identified, which were subdivided into 10 classes: acids, phenols, alcohols, aldehydes, ethers, ketones, hydrocarbons (including aromatic, aliphatic, and acyclic ones), and heterocyclic compounds (Table 6.5).

Individual substances, whose content in the water extracts is no lower than 30 μg/l or 20-80% of the total organic matter in the ultralow-molecular fraction, form the following decreasing sequence: acids, aliphatic hydrocarbons, ethers. The content of phenols, alcohols, aldehydes, ketones, cyclic and aromatic hydrocarbons, and heterocyclic compounds varies within the range 3-10 μg/l.

Different proportions of individual organic substances are observed in the pine forests.

Table 6.5. Composition ol low-molecular organic substances in the water extracts from the litters of the Al-Fe-humus podzols

Parameter	Spruce forest		Pine forest	
	μg/l	% of Σ	μg/l	% of Σ
Σ all fractions	182.8		142.3	
Acids and their Derivatives	86.2	47.1	57.4	40.0
Phenols	5.3	2.9	17.7	12.4
Alcohols	7.5	4.1	13.2	9.3
Aldehydes	3.7	2.1	2.6	1.9
Ethers	19.4	10.6	7.0	4.9
Ketones	2.8	1.6	18.6	13.2
Hydrocarbons				
Aliphatic	39.2	21.5	2.5	1.9
Aromatic	2.5	1.4	7.3	5.3
Cyclic	10.1	5.5	-	-
Heterocycles	6.1	3.2	16.0	11.1
Σ all fractions, % of C_{total}		0.06		0.08

The content of acids, ketones, phenols, and ethers is 12-40 μg/l each (or 10-27% of the total ultralow-molecular organic carbon). Alcohols, aldehydes, heterocyclic organic compounds, and hydrocarbons are present in smaller amounts. Cyclopropane, cyclopentane, heptane, pentalene, benzene, naphthalene, fluorene, and other hydrocarbons were identified.

Volatile organic acids and their derivatives make up a high portion (about half) of the ultralow-molecular WSOSs in the waters of both the spruce and pine biogeocenoses. The absolute content of low-molecular acids in the spruce forest litters is higher than that in the pine forests, which agrees with the differences in the carbon turnover of the biogeocenoses described above.

Low-molecular organic acids (both volatile and nonvolatile ones) in forest litters are mainly inherited from different plant tissues. They are present in the cellular sap of spruce and pine needles (Kaurichev et al., 1991), pectin substances, gums, mucilages, galipots (Soldatenkov, 1971), root exudates of trees and shrubs, and bacterial and fungal metabolites (Aristovskaya, 1965;

Butkevich and Fedorov, 1957; Soldatenkov, 1971; Fox, 1993; Stevenson, 1982).

The content of water-soluble low-molecular organic acids is an integrated result of their influx with falloff and leaching. They have a short lifetime, and their portion in the WSOSs is low (Christe et al., 2000; Huang and Violante, 1986). However, they are involved in podzol formation, as follows from the presence of low-molecular organic acids in the soil solutions of not only forest topsoils but also their Bh horizons (Tani et al., 1993, 1999, 2001). A positive correlation was found between the content of low-molecular organic acids in the soil solutions and their acidity.

Chromato-mass-spectrometric analysis confirmed the presence of phenols in the water extracts from litters; their content in the pine forests (12% of the total ultralow-molecular acid carbon) is significantly higher than in the spruce forests (3%), which agrees with the regularity observed in the Forsyth fractionation of the WSOSs. The spruce and pine forests also differ in other parameters. The water extracts from the spruce forests contain more hydrocarbons and ethers but less alcohols and ketones.

Thus, in the region of the Kola Peninsula studied, which is typical for landscapes at the northern boundary of the taiga zone, a thick litter layer is formed on the surface of the Al-Fe-humus podzols due to the presence of mosses and lichens, the accumulation of plant falloff, and its relatively small alteration under the conditions of low microbiological activity. Litters are a source of water-soluble organic substances and the main zone of their accumulation and transformation.

Organic substances have high level of content in the water extracts from the litters. They mainly include nonspecific low-molecular substances, and low-molecular organic acids are their major components. Functional groups ensure the high chemical activity of the acids. Low-molecular acids and other nonspecific organic compounds like phenols, hydrocarbons, ethers, and ketones, are mainly inherited from plant falloff tissues.

In the litters of northern soils, WSOSs are partly involved in transformation processes, which convert one third of them into substances conventionally classified as fulvic acids. They can differ from the specific organic substances of this group by a lower degree of condensation because of the low microbiological transformation of the organic matter. The content of low-molecular WOSs is quite variable; fulvic acids are thermodynamically more stable substances.

Organomineral Cu and Ni Compounds of Podzols

On the Kola Peninsula, the main soil-forming rocks are bouldered sandy and loamy sandy morainic sediments containing up to 80% of boulders. The mineralogy of the moraine is similar to that of gneisses, granites, and basic intrusions. In the sandy fraction, quartz is predominant (74-80%); finer particles are enriched with accessory minerals inherited from parent rocks, which provide an increased content of many chemical elements (Aleshchukin, 1972; Nikonov and Lukina, 1994; Nikonov and Pereverzev, 1989; Dobrovol'skii, 1972).

It is known (Vinogradov, 1957) that the copper and nickel clarkes are 47 and 58 mg/kg, respectively, in the lithosphere and 20 and 40 mg/kg in soils. The contents of these metals in the regional rocks are 100-150 and 70-80 mg/kg, respectively, which significantly exceed their average content in the lithosphere (Aleshchukin, 1972). Pedogenesis is usually accompanied by the removal of metals from the soil. These trends also remain in sandy and loamy sandy soils of the peninsula: the mineral horizons of podzols contain 10-28 mg/kg of copper and 30-55 mg/kg of nickel. However, the Cu:Ni ratio changes: it is 1.3-2.1 in the soil-forming rocks of the region and decreases to 0.2-0.5 in soils. This can be due to differences in the mobility of the elements and in their retention strength in the primary mineral structure.

The content (mg/kg) of copper and nickel in the studied podzols of the Murmansk region are given in Table 6.6.

In soils of conventional background territories, the content of copper and nickel in litter are 3-7 fold higher than in the mineral horizons. The lowest concentration of metals is in the eluvial horizon. The iron-illuvial horizon acts as a second (after litter) biogeochemical barrier for metals: the total content Cu in it is 2 times higher in comparison with E-horizon. The portion of mobile Cu (1 N CH_3COONH_4 extract) along the whole profile consists of 8-16 %, and for Ni on 2-6 %. Only a part of metals arrived with aerosol waste is retained by the litter : the total Ni content in the litter of technogenic zone prevails 14 times the content of metal in the litter of conventional background territories, Cu – in 27 times. The content of mobile compoundes of metals increased more sufficiently: Ni in 80 times, Cu in 190 times.

The bulk density of litter is sufficiently less (almost one order) then of the mineral horizons, and the thickness of litter decreases on 2-3 times and its acidity increases under the effect of contamination. In such conditions the major part of metals is removed from the litter to the underlying mineral horizons.

Table 6.6. Copper, Nickel, and Their Extractable Compounds in Spruce Forest Podzols of Conventional Background and Technogenic Territories

Horizon, cm	Total content		1 N CH_3COONH_4		H_2O	
	Cu	Ni	Cu	Ni	Cu	Ni
Conventional background area						
A0 0-7	30	137	2.4	3.0	0.4	0.12
E 7-15	6	19	1.0	1.1	0.03	0.02
B_{hfa} 15-23	7	30	0.9	1.3	0.05	0.07
BC 23-38	10	32	1.0	0.8	0.04	0.04
C 38-55	9	35	0.8	0.9	0.04	0.09
Technogenic zone						
A0 0-2	806	1922	462	240	11.7	3.7
E 2-4	16	124	8.7	7	0.26	0.5
Bhfal 4-8	21	137	28.0	19	0.39	1.0
Bhfa2 8-22	35	149	7.3	10	0.05	0.7
BC 22-40	19	139	1.7	9	0.05	0.1
C 40-58	31	121	0.6	2	0.05	0.1

* nd—not determined.

As a result the total content of metals increased in these horizons 3-6 times and the content of mobile metals -7-30 times in comparison with conventional background territories. The composition of lysimetric waters and water extracts from soils is an important indicator of the migration ability of metals (Table 6.7).Podzols of the region have acid reaction (pH 4.5-5.0), which changes to strongly acid (pH < 4.5) under the effect of acid precipitation. Soil waters are ultrafresh in both conventionally background and technogenic zones. The content of Ca^{2+}, Mg^{2+}, K^+, HCO_3^- and Cl^- is low. The content of SO^{2-} ions increases by almost 3 times, compared to the conventionally background soils, under the effect of contamination. The content of water-soluble organic substances decreased in the polluted soils.

Natural waters, including soil solution, are heterogeneous. The content of substances in the true soluble fraction is frequently higher than in colloids and suspensions. For example, they contain 50-95% of total metals in surface, soil, and groundwaters (Dobrovol'skii and Aleshchukin, 1977; Arzhanova, 1977). The diversity of chemical substances in the true solution is due to the interaction of chemical elements with water and dissolved substances during hydration, hydrolysis, ionization, association, and complexation with inorganic (OH^-, Cl^-, SO_4^-) and especially organic ligands.

Table 6.7. Composition of Water Extracts from Spruce Forest Podzols ($\times 10^{-3}$, mmol$_c$/L; C$_{org}$, mg/l)

Horizon, cm		pH	Corg	Ca^{2+}	Mg^{2+}	K^+	HCO_3^-	SO_4^{2-}	Cl^-
Conventionally background area									
A0	0-7	4.64	308	300.0	477	117.8	3.1	124.8	5.3
E	7-15	4.28	104	98.4	16.8	33.3	nd	120.0	nd.
Bhfa	15-23	5.17	58	91.5	19.5	51.3	"	118.2	"
BC	23-38	5.50	nd*	109.3	17.2	32.5	"	125.0	"
C	38-55	5.53	"	92.1	14.9	20.5	"	121.6	"
Technogenic zone									
A0	0-2	3.67	228	0.53	0.09	0.09	0.002	0.32	0.004
E	2-4	4.28	80	0.16	0.02	0.05	nd	0.14	nd
Bhfal	4-8	4.52	46	0.14	0.02	0.04	"	0.11	"
Bhfa2	8-22	4.87	43	0.15	0.03	0.05	"	0.13	"
BC	22-40	5.13	nd	0.29	0.04	0.08	"	0.14	"
C	40-58	5.24	"	0.11	0.03	0.06	"	0.08	"

* nd—not determined.

The active participation of copper and nickel in these reactions resulted in the predominance of bound metal compounds differing in composition, molecular weight, and charge in water extracts from litters of northern podzols. The relative content of free copper ions is no more than 10-13% of the total metal in the extracts. Ladonin et al. (1994) showed that the activity of Cu^{2+} ions in solutions of natural and artificially contaminated soddy-podzolic soils corresponded to pCu 5.1-7.2 or 1-30% of their total content in solution. Similar contents of free metal ions are frequently observed in surface waters (Hodson et al., 1965). For example, the content of Cu^{2+} and Ni^{2+} ions in water of the Moskva and Kura rivers makes up 13-29% of their total amount (Supatashvili and Makharadze, 1988).

As early as 1936, Vernadskii asserted that chemical elements in natural waters predominantly occur as complex ions rather than as simple ions. Aleksandrova (1980) and Kononova (1951) also concluded that only a small portion of organic substances occurs in soils and waters in the free state; most organic substances are bound to mineral components. As the methods of study were improved, these conclusions were confirmed by experimental data on the material composition of the true soluble fraction of natural water.

In the soils studied, the WSOSs content varies from 100 to 308 mg/l (0.1-30 mM/l); i.e., it exceeds the content of metals (n x 10^{-3} mM/l) by 3-4 orders of magnitude. The total content of WSOSs and associated metals decreases with depth. The predominance of WSOSs results in the prevalence of organomineral compounds of bound metals in liquid phases. The capacity of WSOSs to actively interact with metals was already noted in 1940s-1950s (Rode, 1941; Skrynnikova, 1959; Tyurin, 1944; Khan, 1951).

In the first studies, attention was focused on organomineral compounds of iron and copper (Zhuravleva, 1978; Kaurichev and Nozdrunova, 1961; Manskaya et al., 1958). Further study of organomineral compounds of metals in natural waters is related to the development of analytical techniques (Varshal et al., 1993; Kaurichev et al, 2004; Blaser et al., 1984; Guggenberger et al., 1989; Schulten, 1999; Titeux et al., 1999).

The ecological significance of WSOSs is related to their various interactions with metal ions. Information on the composition of WSOSs and associated metals in water extracts from the northern taiga podzols studied was derived from their fractionation using the Forsyth method. Organic and organomineral substances were sorbed on activated carbon and then successively desorbed with 0.1 N HCl, water-acetone solution, and 0.5 N NaOH (Table 6.8).

Table 6.8. The content of C_{org}, Cu, and Ni in water extracts of the litters from the soils of technogenic zone and their compounds after the fractionating according the method of Forsyth

Element	Biogeocenosis	Initial water extract, mg/l	organic and organomineral substances (% from total content in extracts)		
			nonspecific organic compounds		fulvic acids
			0.1 N HCl	acetone-water	0.5 N NaOH
C_{org}	spruce forest	100-146	10-14	44-64	14-21
	pine forest	59-95	42-25	46-73	28-45
Cu	spruce forest	1.4-1.7	48-59	25-30	9-10
	pine forest	0.5-0.6	51-57	28-32	20-23
Ni	spruce forest	0.4-0.6	34-44	27-35	16-21
	pine forest	0.2-0.3	42-50	27-32	31-36

The fraction of solutions from litters hypothetically classified among fulvic acids contains, in most cases, no more than 1/3 of organic substances. Their low concentration is explained by the relatively low degree of transformation of organic remains under northern taiga conditions. However, this fraction is characterized by the best agreement between the contents of metals and organic substances. Fulvic acids are considered as the most active complexing agents in surface waters, where they can bind up to 90% of the metals present (Varshal et al., 1979, 1993; Makharadze, 1988).

Chromatographic analysis of water extracts from litters of the soils studied revealed the presence of phenols; oxalic, lactic, propionic, and succinic acids; and numerous volatile organic acids. These substances are inherited by litters of taiga soils from plant residues, needles, and falloff (Fox and Comerford, 1990; Tani et al., 2001). Carbohydrates actively consumed by microorganisms are less stable in soils; phenols are relatively stable compounds with numerous functional groups (Kaurichev et al., 1991).

Among soluble organic substances in podzol litters of the Kola Peninsula, phenols are presumably predominant (44-73% of total C_{org} in the extract). The portion of copper and nickel in this fraction is almost 2 times lower (about 30% of the total content in the extract). This disagreement can be due both to the involvement of phenols into interaction with other metal ions, whose complexes exceed those of copper and nickel in amount and stability (e. g., with iron ions), and to the occurrence of some phenols in the free state.

About half of soluble metals were found in the fraction presumably bound to organic acids, purine bases, and monosaccharides, which make up no more than 1/4 of total organic substances in the extract. Metals were probably bound to these substances with the formation of chelate complexes and relatively unstable heteropolar salts resulting from the exchange of metals for carboxyl and phenolhydroxyl protons (Aleksandrova, 1980). This fraction can also contain hydroxo complexes of metals, mixed hydroxo-organic complexes, and free copper ions. The absolute content of metals and organic substances (total and in fractions) in the extracts from litters of spruce forests is almost double that of pine forests, which is related to differences in the falloff composition and its decomposition conditions. However, the relative contents of the separated fractions are almost similar in two biogeocenoses.

The capacity of water-soluble organic substances to bind metals is due to the presence of various structural fragments and functional groups (C=O, C-OH, C-OOH, C=C, $=N^-$, S-H, and others) capable to form, along with heteropolar salts, coordination bonds with metals having incomplete *d*-electron subshells. In the resulting chelate compounds, metals enter into the anionic part of the molecule and are incapable of exchange reactions (Varshal et al., 1979; Stevenson and Fitch, 1981). Varshal et al. (1979) think that fulvic acids are predominant in natural waters and that their monomers are similar in nature to monohydroxybenzene-di- and tricarboxylic acids and dihydroxybenzene-di- and tricarboxylic acids; i.e., monomers of fulvic acids contain two-three carboxyl groups and one-two phenolic hydroxyls; these functional groups ensure a strong bond between the ligand and metal ions. It is suggested that the functional groups of salicylic and phthalic acids inherited from plant residues actively participate in complexation (Moore and Ramamoorthy, 1987).

As for metals, their capacity for complexation with organic substances in natural waters is directly related to their ionic potential (Krainov, 1973). It is assumed that this capacity is maximum in metals with an ionic potential of > 2.5. Among them are copper and nickel; therefore, their organomineral complexes prevail in natural waters (Supatashvili and Makharadze, 1988). The absorption of metals depends on the properties of WSOSs, primarily on the nature of functional groups, their number, and the reaction of the environment.

The molecular-weight distribution of organic substances and organomineral compounds of copper and nickel also points to the insignificant participation of high-molecular organic substances in the interaction with soluble metal compounds (Table 6.9).

Table 6.9. The content of C_{org}, Cu, and Ni in the fractions with different molecular weights in water extracts from the litters of technogenic zone

Element	Total content, mg/l		The portion in the fractions (% of total)							
			> 20000		19300-19000		2200-1700		1700-1200	
			Da							
	pine forest	spruce	pine	spruce	pine	spruce	pine	spruce	pine	spruce
C_{org}	59-95	100-146	10-20	34-45	3-15	22-31	52-70	10-20	12-17	10-17
Cu	0.53-0.60	1.38-1.69	12-18	12-14	9-24	18-22	30-46	29-33	18-37	25-31
Ni	0.22-0.26	0.43-0.56	16-22	20-26	3-7	10-16	29-49	38-40	36-41	20-31

Organic substances in the extracts from litters of northern taiga podzols are characterized by molecular-weight heterogeneity. Among them, two predominant groups are separated: high-molecular substances with MW > 19000 Da and relatively low-molecular substances with MW < 2200 Da. The portion of low-molecular substances is double that of high-molecular substances (60-71 and 31-34%, respectively). They are conventionally classified among nonspecific organic substances. Organomineral compounds of copper and nickel are concentrated in this fraction (48-90%), which indicates the highest activity of WSOSs with MW < 2200 Da in complexation with metals. However, the contributions of separate groups of individual organic substances (nonspecific substances, fulvic acids, hydroxo complexes, and hydroxo-fulvate complexes) to the formation of organomineral compounds of metals could not be determined in this case, because a relatively wide MW range of the predominant fraction was distinguished in chromatograms, which could be due to the presence of many of these substances. The literature data indicate that more split fractionation of natural waters could reveal the bonds of metals with specific groups of soluble organic substances (Varshal et al., 1979, 1993; Makharadze, 1988).

Varshal (1993) assign the leading role in the formation of migration-capable compounds of metals to fulvic acids presumably predominant (60-80%) in surface waters. Their content is 3-33 mg/l on average and sometimes exceeds this level by 3-4 times. Fulvic acids are considered to be polyfunctional polydispersed complexing substances susceptible to association in water solutions. This explains the wide range of pH-dependent molecular weights of substances presumably classified as fulvic acids: from 300 (monomers) to 60000 (associates). For example, it was found for surface waters with pH 3.25 that the molecular weight of the FA monomer is 273-286 (Varshal et al., 1979; Makharadze, 1988). These groups of WSOSs bind metals in surface waters almost completely. For example, it was found for river waters from Georgia that the major contribution to the budget of organomineral compounds of copper was from two groups of substances, with MW > 1000 Da and MW < 180 Da (Makharadze, 1988).

The organomineral particles of Cu and Ni compounds in water extracts have different charge (positive, negative), part of them are uncharged. There is a lot of information about the relation of these particles in natural water. Among bound metals in positively charged particles, not only organomineral compounds of metals are apparently present, but also their hydroxo complexes of type $Cu(OH)^+$ and, possibly, hydroxo-organic complexes. The anionic and neutral fractions can include simple hydroxo complexes of types $Cu(OH)_2^0$

and mixed compounds (hydroxo-organic complexes of type $[Cu(OH)Lig_{org}]^-$). For example, the formation of complex $[Cu(OH)_2(FA)_m]^{2m-}$ with stability constant $\beta_{1,1} = (3.08 \pm 2.01) \times 10^6$ is most probable in the presence of fulvic acids, and a consistent increase in the concentration of copper ions leads to the predominant formation of complexes $[Cu_4(FA)_2]$ (Varshal et al., 1979; Schnitzer and Kerndorf, 1981).

The literature data attest that the predominance of organomineral compounds of metals with zero and negative charges is most typical for water-soluble compounds of Fe, Al, Cu, Zn, and Pb in acid soils (Dobrovol'skii and Aleshchukin, 1977; Aleshchukin, 1990, 1992).

For example, it was found that these forms prevail in liquid phases of brown acid tundra, podzolized tundra, and peat-bog-gley soils from the Malozemel'skaya tundra, thick ordinary chernozems from the Dnepropetrovsk region, and krasnozems from Adjaria. Their portion frequently decreases with depth (Aleshchukin, 1992).

In organic horizons of tropical island soils in the Indian Ocean, most organomineral iron compounds (39-59%) occur in the electrically neutral form, and anions make up 25-40% of true soluble metal compounds. The predominance of neutral complexes (45%) and anionic forms (36%) was also revealed for zinc; anionic forms prevailed for copper and lead (up to 47 and 50-38%, respectively) (Dobrovol'skii, 1990).

The relative contents of uncharged and negatively charged organomineral metal compounds in natural waters are frequently similar; the predominance of one group over the other is indistinct. For example, anionic zinc forms slightly prevail in the A0 horizon of tundra soils (51%), and neutral particles are predominant in the A_0/A_T horizon (45%). An inverse relationship is observed in peat-bog soil, which is similar in properties (Aleshchukin, 1992). However, uncharged and negatively charged complexes are characterized by a high migration capacity.

Thus, the migration capacity of copper and nickel arrived with aerosol waste in the profile of regional podzols is due not only to the percolative water regime of soils, but also to the presence of water-soluble organic substances in the litter. The predominant substances by 2/3 consist of low-molecular nonspecific organic substances and by 1/3 of newly formed substances conventionally classified as fulvic acids. Most metals are in the bound state. Fulvic acids are actively bound by metals; there is a proportionality between their contents in the resulting complexes. Low-molecular organic substances, among which prevail acids and phenols predominantly inherited by litters from plant residues, are the most active in the interaction with copper and nickel.

Correlation of metals with organic substances is less close than with fulvic acids, which is explained by the diversity of the resulting chelates and by the presence of metals in hydroxo complexes and hydroxo-organic complexes. Part of the metals can be loosely bound to organic substances such as heteropolar salts, where ions of some metals can be replaced by ions of other metals. Sufficient part of organomineral metal compounds can occur as neutral or negatively charged particles with a high migration capacity. The predominance of these compounds in the solution results in the depletion of metals in soil litters of natural landscapes and in the reduction of metal accumulation in the litters of contaminated soils. Migration-capable organomineral metal compounds are not only transported by water flows to the lower part of the soil profile, but are also removed beyond its limits. The ecological significance of these processes is that they result in the increase in the concentrations of copper and nickel in the groundwaters of the technogenic zone of the region, which in some cases exceed the MPC level for potable water (Motuzova et al., 2004).

Chapter 7

COMPLEXATION OF METAL IONS WITH ORGANIC SUBSTANCES AND THE ACID–BASE PROPERTIES OF THE SOIL LIQUID PHASE IN THE TAIGA AND STEPPE ZONES

Complexation of metal ions with soil organic substances is one of the main mechanisms for the dissolution, transfer, and concentration of metals in natural environments. These processes are involved in the weathering of minerals of soil-forming rocks, the transformation of chemical compounds present in them, and their subsequent migration and accumulation. They largely regulate the behavior of metals in natural waters. In soil solutions and lysimetric and surface waters, 50 to almost 100% of metals occur in the form of complexes (Varshal et al., 1983; Linnink and Nabivanets, 1986; van Hees et al., 2000; Essington, 2002; Linnink and Vasilchuk, 2005). The complexation of metals with soil organic substances plays an important role in plant nutrition: it decreases the hazard of metal excess for living organisms, because free metal ions are the most toxic (Santore et al., 2001; Inada, Takenaka, 2005). The interaction of metal ions with water-soluble organic substances (WSOSs) favors the formation of mobile metal forms necessary for plants and increases their availability to plants under deficit conditions (Lundstrom et al., 2000; Kachl and Chen, 2005).

Soil solutions contain various dissociated and nondissociated organic substances capable of interacting with metal ions: low-molecular-weight organic acids (LMWOAs), amino acids, carbohydrates, and humic substances (HSs). Differently stable complexes of these substances with iron, aluminum, manganese, and other metals are also present in soil solutions. The content and stability of complexes in the soil liquid phase depend on internal and external factors.

Water-soluble organic substances are essential sources of the soil-solution acidity and have the leading effect on the development of chemical properties of the entire soil. Their participation in complexation with metals disturbs the dynamic acid–base equilibrium in solutions, because metal ions compete with protons for the binding sites of organic substances. The displacement of protons from the functional groups of organic substances by metal ions and the complexation of metals with organic ligands are accompanied by the acidification of solution. The released protons partially enter the reverse reaction of protonation, which results in the establishment of a dynamic equilibrium determining the acid–base properties of the solution. However, the relationship between the complexation of metal ions with organic substances and the acid–base properties of soil liquid phase is insufficiently understood. This determined the aim of the investigation. In this Chapter, the mutual effects of the complexation of metal ions with organic substances and the acid–base properties of the liquid phase in soils of the taiga and steppe zones will be shown.

We studied water extracts from the organic and mineral horizons of soils in the northern and southern taiga and steppe zones of the Russian Federation: the F litter subhorizon of podzolic soil, Kola Peninsula, Murmansk oblast; the Ad horizon of soddy soil, Konakovo district of Tver oblast; the H litter subhorizon of bog-podzolic soil and the F litter subhorizon and A1A2 horizon of podzolic soil, Central Forest State Nature Biospheric Reserve; and the Ap horizon of chernozem, Voronezh oblast.

The main method used for solving the set problem was the potentiometric titration of water extracts from the soils studied with metal salt solutions, predominantly copper nitrate, using glass and copper-selective electrodes. From the titration curves, the activity of ions can be determined as a function of the titrant volume, the point of equivalence can be found, and information can be acquired on the state (free, bound) of protons and copper ions. This method of analyzing water extracts from soils is described in Chapter 3 (6). It involves the combined potentiometric acid–base and complexometric titration with a weak acid, because copper ions, when added to water extracts from soils, manifest properties of a weak acid due to hydrolysis and properties of a strong complexing agent reacting with WSOSs. Both processes are accompanied by the release of additional protons. The simultaneous recording of the content of copper ions and protons varying during the titration allows the correlation between the complexation of copper ions with WSOSs and the acid–base properties of water extracts from soils to be revealed.

Acidification can be due not only to the above processes, but also to the displacement of iron and aluminum ions from the present complexes with WSOSs and their subsequent hydrolysis. Related information can be acquired from the assessment of the content of free Fe^{3+} and Al^{3+} ions in the analyzed extracts after the addition of copper salts, as was done in this work.

Water extracts from the studied soils strongly differed in their properties. The extract from the H litter subhorizon of bog-podzolic soil had an acid reaction (pH_{H2O} 4.2), relatively high content of C_{org} (112 mg/l) and total acidity (2.3 $mmol_c$/l), and insignificant contents of alkali metals and trace elements (Cu, Zn). The extracts from the upper horizons of chernozem and soddy soil had near-neutral pH; their acidities were lower by two times and the concentrations of alkali and alkali-earth metal higher by an order of magnitude compared to the extract from bog-podzolic soil (Table 7.1).

Table 7.1. Chemical properties of water extracts from the soils studied

Horizon	pH	C_{org}, mg/l	Total acidity, $mmol_c$/l	Cation concentrations, $mmol_c$/l						HCO_3, $mmol_c$/l
				Ca^{2+}	Mg^{2+}	K^{+}	Fe^{3+}	Cu^{2+}	Zn^{2+}	
Podzolic soil, Kola Peninsula										
F	4.6	26	nd	0.30	0.04	0.11	nd	0.0006	0.0002	nd
Podzolic soil, Central Forest Reserve										
F	5.2	64	nd	0.41	0.07	0.12	0.030	0.001	0.003	nd
A1A2	5.5	31	nd	0.19	0.03	0.59	0.010	0.001	0.002	nd
Bog-podzolic soil										
H	4.2	112	2.3	0.08	0.02	0.07	0.070	0.0005	0.001	nd
Soddy soil										
Ad	6.4	45	1.2	0.76	0.24	0.64	0.030	0.001	0.002	0.35
Chernozem										
Ap	7.1	165	nd	0.81	0.21	nd	nd	0.001	0.002	0.50

Note: (nd) not detected.

The extracts contained different organic acids, as was shown by chromatographic study (Table 7.2). The found acids differed in strength: their pK_a values were 3.75 and 4.76 for formic and acetic acids, respectively; pK_{a1} and pK_{a2} for oxalic acid were 1.27 and 4.27, respectively; pK_{a1}, pK_{a2}, and pK_{a3} for citric acid were 3.13, 4.76, and 6.40, respectively (Lindsay, 1979; Lur'e, 1989). The pH values in the Ad and Ap horizons were close to 7; therefore, most acids were in the dissociated state.

Table 7.2. Concentrations of organic acid anions in the extracts

Horizon	Formiate, mmol/l	Acetate, mmol/l	Oxalate, mmol/l	Citrate, mmol/l
Ad	nd	1.453	0.076	0.001
H	0.408	0.703	0.151	0.001
Ap	0.106	1.709	0.066	nd

Note: (nd) not detected.

According to literature data (van Hees et al., 1996; Lundstrom et al., 2000), the concentration of LMWOAs in soil solutions of podzolic soils varies in a wide range: 0.001–1.8, 0.001–0.450, 0.001–0.080, 0.001–0.370 mmol/l for the acetate, formiate, oxalate, and citrate ions, respectively. The upper limits correspond to the LMWOA concentrations in the solutions of organic soil horizons, and the lower limits correspond to their concentrations in solutions of the B horizons. In the extracts from the upper soil horizons under study, high LMWOA concentrations were observed, except citric acid, which was present in lower concentrations in the extracts from the Ad horizons of soddy soil and the H subhorizon of bog-podzolic soil and was not found in the extract from the Ap horizon of chernozem. The extract from the H subhorizon was characterized by an increased content of oxalates.

The results of continuous potentiometric titration showed that the extracts contained substances ensuring their acid–base buffer properties in a wide pH range. The highest contents of buffer components were titrated in the pH ranges 3.0–4.0 and 9.0–10.0, as is usually observed for the acid–base buffer capacity of extracts and soil solutions (Ivanova and Sokolova, 1998; Sokolova et al., 2007). The substances reacting in the pH range 3–7 can be classified as carboxyl groups, and those dissociating in the pH range 7–10 as phenolhydroxyl groups.

Parameters of Complexation of Copper Ions with Oxalate Ions

At the first stage, model solutions of organic substances of known composition were titrated with a copper salt solution. Oxalic acid solutions of different concentrations (0.65 and 0.75 mmol/l) were used as model solutions, because this compound is a permanent component of WSOSs in soils of the taiga zone (Kaurichev et al., 1963; van Hees et al., 2000; Motuzova et al., 2003), which can form stable complexes with metal ions (Lur'e, 1989).

Table 7.3. Parameters of complexation of copper ions with oxalate ions

C ($H_2C_2O_4$), mmol/l	M_{max}, mmol/l (Gran)	lgK (Henderson–Hasselbalch)	M_{max}, mmol/l (Langmuir)	lgK (Langmuir)	Cu/C_{org}	H/Cu
0.65	0.52	5.9	0.54	5.6	0.4	1.0
0.75	0.59	6.0	0.57	5.6	0.4	1.2

The parameters of complexation of copper ions with oxalate ions calculated from the experimental data were identical to those obtained using the Gran method and Langmuir equation, which indicated the information values of both calculation procedures (Table 7.3).

During the titration of oxalic acid solutions (0.65 and 0.75 mmol/l) with a copper salt solution (0.013 mol/l), the major part, although not all, of oxalate ions (0.52–0.59 mmol/l) was capable of complexing with copper ions in the solution. The complexation was hampered by the decrease in pH from 3.0–3.2 (which is typical for an oxalic acid solution) to 2.9 during the titration because of the displacement of protons from the functional groups of oxalic acids by copper ions: $Cu^{2+} + HC_2O_4^- = CuC_2O_4^0 + H^+$. The additional protonation increased the competition of protons for the active sites of the oxalic acid dissociation products, which restricted the complexation with their participation.

The stability constants of copper complexes with oxalate ions calculated from the Henderson–Hasselbalch equation were identical to those calculated from the Langmuir equation; the lgK values of the complexes were in the range 5.5–6.0 (Table 7.3). Their calculation was based on the assumption that the 1 : 1 complexes were formed.

The Cu/C_{org} value, which represents the ratio of the complexed copper ions (M_{max}, mmol/l) to the concentration of organic carbon (mmol/l), can characterize the ability of organic substance to form complexes with metal ions. The value obtained for the oxalate complexes of copper (0.4) indicates a high ability of oxalic acid to complexation with cooper ions.

From the data on pH changes, the number of moles of protons released at the binding of one mole of copper ions (H/Cu) was calculated. The obtained data indicated that one proton, on the average, was released into solution at the binding of one copper ion.

The changes in the concentrations of free Cu^{2+} ions, those complexed with oxalate ions, and protons released during the titration of oxalic acid with a copper nitrate solution occurred consistently. The content of copper in oxalate complexes increased; therefore, free copper ions were not observed in the solution at the initial stages, but they appeared in the course of titration, and the content of protons increased (Figure 7.1).

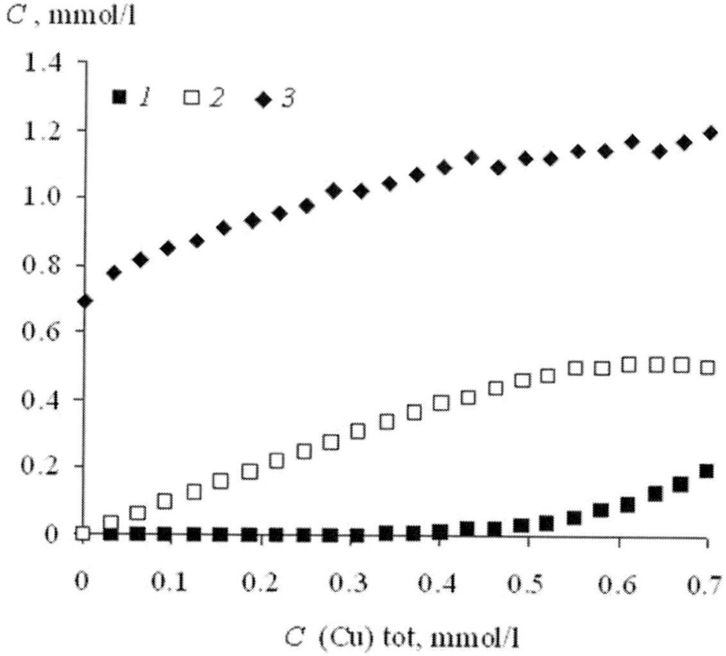

Figure 7.1. Concentrations of (1) free Cu^{2+} ions, (2) those complexed with oxalate ions, and (3) protons as functions of the total concentration of copper ions added during the titration of a 0.65 mmol/l oxalic acid solution with a copper nitrate solution.

Parameters of complexation of copper ions with organic substances of water extracts from the studied soils. The tendencies observed in the course of the potentiometric titration of water extracts with a copper salt solution were similar to those observed during the titration of model oxalic acid solutions. At the first stages of titration, accumulation of bound copper forms occurred in the solution; free copper ions were absent, and the content of protons regularly increased. Acidification of solutions because of the complexation of metal ions with organic substances was noted for different preparations of HSs (Kaemmerer et al., 1999; Robertson, Leckie, 1999; Osterberg et al., 1999). At the next stage of titration, the accumulation of bound metal forms gradually ceased, whereupon the content of free copper ions in the solution increased linearly.

Parameters calculated from the Langmuir equation and the Gran function, which largely characterize all ligands present in the extract, are given in Table 7.4. The studied extracts differed in complexation parameters.

Table 7.4. Parameters of complexation of WSOSs in the studied soils with copper ions

Horizon	M_{max}, mmol/l (Langmuir)	*lgK* (Langmuir)	M_{max}, mmol/l (Gran)	*lgK* (Henderson–Hasselbalch)	Cu/C_{org}	H/Cu
Ad	0.23	5.6	0.23	5.7	0.03	0.1
H	0.30	3.9	0.27	4.3	0.015	0.5
Ap	0.55	5.0	0.56	5.8	0.04	0.1

The complexing capacities of WSOSs for copper ions in extracts from the studied soils differed by no more than 2.5 times. Data are available on the wide variation of the complexing capacity of WSOSs in natural waters (from 1 nmol/l to 1 mmol/l) depending on the content and quality of dissolved substances (Town and Filella, 2000). Among the soils under study, WSOSs from the Ap horizon of chernozem had the highest complexing capacity for copper ions (0.55 mmol/l). This was favored by the high content of organic matter in the chernozem extract (165 mg/l) and the favorable extract reaction (pH 7.1). WSOSs from the Ad horizon of soddy soil and the litter horizon of podzolic soil were similar in complexing capacity for copper ions, which was almost half that in the extract from the upper horizon of chernozem.

However, these objects had different sources of this feature, in accordance with significant differences in their properties (reaction, the content of WSOSs, the composition of WSOSs). The highest stability of copper complexes with WSOSs was also noted for the extracts from the Ad horizon of soddy soil and the Ap horizon of chernozem (*lg K* 5.0–5.6). The near-neutral reaction of these soils was the leading factor for the formation of stable complexes with copper ions. The formation of complexes was accompanied by an increase in acidity of the titrated solutions.

The maximum Cu/C_{org} value was observed for the chernozem WSOSs, which agreed with the high values of complexing capacity and stability constants of copper complexes with WSOSs. The minimum value was found for the WSOSs from bog-podzolic soil because of its low pH unfavorable for the complexation of metal ions with organic substances.

The number of protons released at the binding of WSOSs by copper ions (H/Cu) depended on the pH level. In extracts from the Ad horizon of soddy soil (pH 6.3) and the Ap horizon of chernozem (pH 7.1), the formation of WSOS complex with one mole of copper resulted in the release of 0.1 mole of protons. The acid-buffer capacity of extracts from these soils was high, and the yield of protons was low because of the near-neutral reaction of these solutions and the presence of organic acids in dissociated state. In the extract from the H subhorizon of bog-podzolic soil (pH 4.6), the complexation of one mole of copper ions released 0.5 mole of protons. The effect of pH on the content of protons released at the complexation of metals with HSs was reported in the literature. It was noted that the complexation of one copper ion with humic acids (HAs) entailed the release of less than one proton into solution at pH 7 and 2.8–3.9 protons at pH 4 (Osterberg et al., 1999). In another case, the interaction of HAs with one copper ion released 1.3–1.4 protons at pH 4 and 0.8–1.0 protons at pH 6 (Robertson, Leckie, 1999). The effect can also depend on the content of added copper ions, the concentration of HAs in the solution, the composition of organic substances, the number of functional groups involved in the binding of copper ions, and the degree of ionization of organic acids.

The complexation parameters calculated using the above methods characterize WSOSs in the studied extracts as a whole, i.e., they represent average characteristics of ligands present in the extracts. The extract WSOSs form a continuum of ligands; each of them is characterized by specific values of Me_{max} and K (Fitch et al., 1986; Town and Filella, 2000). To characterize the diversity of ligands present in the extracts in more detail, the Scatchard method was used, which allows distinguishing two ligand types (Saar and Weber, 1979; Kolosov, 1982; Fitch et al., 1986; Wu and Tanoe, 2001).

The plotting of graph for the complexation of copper ions with WSOSs of extracts from the upper horizon of soddy soil in the Scatchard coordinates revealed the participation of two ligand types in complexation in the extracts from soddy soil and chernozem and allowed calculating the corresponding complexation parameters (Table 7.5). In the extract from Ad horizon of soddy soil, an insignificant amount of ligands of the first group participated in the complexation with copper ions (no more than 0.06 mmol/l).

They were characterized by an increased affinity for copper ions (lgK 7.3) and could include the most active phenol hydroxyls detected during the continuous potentiometric titration of extracts in the pH range 7–9. Model calculations (Benedetti et al., 1995) showed that phenol hydroxyls could prevail in the binding of copper ions in the presence of low copper concentrations even at pH 6.

Table 7.5. Complexation parameters of WSOSs calculated by the Scatchard method

Horizon	M_{max} 1, mmol/l	*lgK* 1	M_{max} 2, mmol/l	*lgK* 2
Ad	0.06	7.3	0.19	5.0
H	nd	nd	0.28	3.8
Ap	0.10	8.7	0.52	3.7

Note: (nd) not detected.

As far as the strong ligands were consumed, the concentration of the added copper ions increased, and the pH decreased, the weaker ligands of the second type entered into the reaction. Their complexes with copper ions were less stable (*lgK* 4.3), but they bound more metal (0.19 mmol/l). It can be believed that the ligands of the second type largely characterize the organic substances prevailing in the extracts (Fitch et al., 1986).

In WSOSs of the extract from the Ap horizon of chernozem, ligands of two types were also identified. The regularities of their complexation with copper ions were similar to those described above; the ligands of the first group (0.10 mmol/l) less actively participated in complexation with copper ions than the ligands of the second group (0.52 mmol/l). The stability of complexes formed with the participation of ligands of the first group was higher compared to the complexes with ligands of the second group (*lgK* 3.7). The decrease in the stability of complexes of copper ions with the WSOS ligands of the second group was due to the acidification of solutions at the formation of these complexes. Data are available that the stability constants of metal complexes with natural ligands can decrease even by orders of magnitude with decreasing pH (Cheam and Gamble, 1974; Saar and Weber, 1979).

For the extract from the H subhorizon of bog-podzolic soil, the graphical splitting of its titration plot with copper salt revealed no different types of WSOS ligands. Under acid conditions of this soil, a lower diversity of functional groups capable of interacting, with the predominance of carboxyl groups, was observed.

To increase the complexation of copper ions with WSOSs at the potentiometric titration of extracts from the Ad horizon of soddy soil and the H subhorizon of bog-podzolic soil with a copper salt solution, an additional amount of oxalic acid (1.6 mmol/l) was added to the extracts. Changes in the proportions of free and bound copper forms and the acid–base conditions of extracts were analyzed (Figure 7.2).

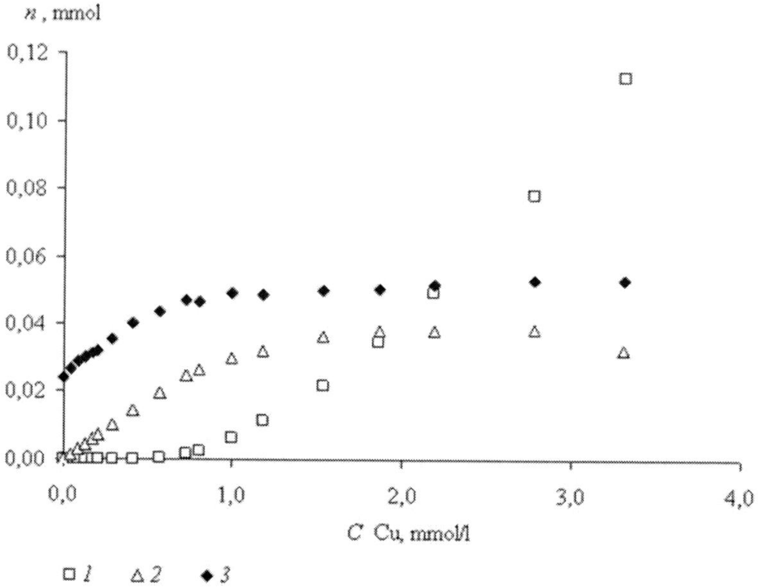

Figure 7.2. The contents of free and bound copper and protons as functions of the concentration of added copper ions during the titration of the extract from the Ad horizon in the presence of oxalic acid.

Table 7.6. Parameters of complexation of the initial extracts and the extracts with the addition of oxalic acid

Object	pH_{ini}	Me_{max}, mmol/l (Langmuir)	*lgK* (Langmuir)	Cu/C_{org}
H, bog-podzolic soil	4.42	0.30	3.9	0.015
Ad, soddy soil	6.34	0.23	5.7	0.03
H, bog-podzolic soil + additive	3.00	1.16	4.8	0.05
Ad, soddy soil + additive	3.15	1.29	4.8	0.09

After the addition of oxalic acid, the extracts became more similar in pH and parameters of complexation with copper ions. The complexing capacity of WSOSs for copper ions in the extracts was enhanced; the Cu/C_{org} ratio in the complexes increased, as well as the content of protons released during the complexation. The constant of complexation decreased in the extract from the Ad horizon and increased in the extract from the H subhorizon.

Effect of metals on the acid–base properties of water extracts from soils. The decrease in the pH of water extracts from soils with different properties during their titration with salt solutions of different metals (Fe, Al, Pb, Zn, Hg) was confirmed (Table 7.7).

Table 7.7. Effect of the titration of water extracts from soils with different properties by metal salt solutions on the acidity of extracts

Salt	0.01 M salt solution pH	Extract pH	TEP pH
F subhorizon, podzol			
$Fe_2(SO_4)_3$	2.31	4.14	2.44
$Cu(NO_3)_2$	5.00	4.22	4.05
$Zn(NO_3)_2$	4.88	4.14	3.75
$Pb(NO_3)_2$	4.89	4.14	3.64
F subhorizon, podzolic soil			
$Fe_2(SO_4)_3$	2.31	4.36	2.48
$Cu(NO_3)_2$	5.00	4.13	4.08
$Zn(NO_3)_2$	4.88	4.36	3.88
$Pb(NO_3)_2$	4.89	4.36	3.64
A1A2 horizon, podzolic soil			
$Fe_2(SO_4)_3$	2.31	4.53	2.48
$Cu(NO_3)_2$	5.00	4.53	4.32
$Zn(NO_3)_2$	4.88	4.53	4.24
$Pb(NO_3)_2$	4.89	4.53	4.18
H subhorizon, bog-podzolic soil			
$Cu(NO_3)_2$	5.00	4.42	3.94
$Pb(NO_3)_2$	4.89	4.42	3.92
$Al(NO_3)_3$	4.17	4.42	3.80
$HgCl_2$	4.44	4.40	4.33
Ad horizon, soddy soil			
$Cu(NO_3)_2$	5.00	6.34	5.10
$Pb(NO_3)_2$	4.89	6.34	5.39
$Al(NO_3)_3$	4.17	6.34	4.51
$HgCl_2$	4.44	6.36	5.77
$Hg(NO_3)_2$	2.84	6.31	2.71
horizon Ap, chernozem			
$Cu(NO_3)_2$	5.00	7.09	5.77

The effect depended on the properties of both extracts and metals. The strongest acidifying effect was observed for the ions of Fe, Hg (added as nitrates), and Al. These metals are not only capable of complexing with WSOSs, but they also manifest relatively strong acid properties because of hydrolysis. The ions of Zn, Pb, and Cu have relatively low abilities of hydrolysis, but they are also capable to significantly decrease the pH level even of initially acid extracts because of their complexation with WSOSs. Among the studied salts, the lowest acidifying effect was observed for the Hg^{2+} ions added as chlorides. Chloride ions can form stable complexes with Hg^{2+} ions; therefore, the capacity of Hg^{2+} ions for the complexation with WSOSs and for hydrolysis is restricted.

Among the extracts from different soils, the highest acidifying effect was observed in the extracts from soddy soil and chernozem because of their near-neutral reactions. This reaction favored both the complexation with WSOSs and hydrolysis; therefore, the most significant decrease in pH was observed for these extracts. In the extracts with the initially acid reaction, the acidifying effect was less pronounced, because the acid reaction favored neither complexation nor hydrolysis. In addition, the acid-buffer capacity of extracts increased in the region of low pH.

Release of aluminum and iron. In soil solutions, organic acids are present not only in the protonated or dissociated form, but also as complexes with such metals as iron and aluminum. In the course of interaction between copper ions and organic substances of the extracts from the Ad horizon and the H subhorizon, the release of iron and aluminum from the bound state was revealed (Table 7.8). The mechanism of this process can be described as follows:

$$AlL^{+} + Cu^{2+} = CuL + Al^{3+}.$$

The released iron and aluminum ions are then subjected to hydrolysis, which results in an increase in the activity of protons in the solution:

$$Al^{3+} + H_2O = AlOH^{2+} + H^{+}.$$

Thus, the participation of aluminum and iron contributes to changes in the acid–base equilibrium and can increase the heterogeneity of binding sites, because iron and aluminum differently compete with copper for binding sites.

Table 7.8. Changes in the concentrations of free aluminum and iron ions in extracts under the effect of copper ions

Sample	*C Al*, mmol/l	*C Fe*, mmol/l
Extract from horizon Ad	0.002	0.030
Extract from horizon Ad + 0.42 mM Cu^{2+}	0.007	0.048
Extract from subhorizon H	0.201	0.070
Extract from subhorizon H + 0.42 mM Cu^{2+}	0.362	0.082

The potentiometric titration of water extracts from some soils with a copper salt solution at the successive control of the activity of protons and copper ions revealed relationship between the complexation of metal ions and organic substances of water extracts from soils with different features and the acid–base properties of the extracts. It was found that the addition of increasing amounts of metals to water extracts from the soils containing WSOSs resulted in the successive formation of at least two types of complex compounds. From the data on the complicated composition of soil WSOSs, it can be concluded that the number of such types can be larger, and they can form a continual series of complex compounds, some of which cannot be detected in the course of titration. During the interaction of Me (in this case, copper) ions and WSOSs at the low initial Me/ligand ratio, the most active WSOS components capable of forming more stable complexes with Me enter into the reaction with Me. The capacity of forming such complexes is relatively low (0.05–0.10 mmol/l). The complexation is accompanied by the acidification of solution under the effect of some processes, whose contributions depend on the properties of both metals and WSOSs. These processes include the acid hydrolysis of metal ions added to the solution (Cu, Zn, Pb, etc.), the displacement of protons from the functional groups of organic acids by protons, the complexation of metal ions with organic acid ligands with the attendant release of protons, the substitution of iron and aluminum ions in the present complexes with WSOSs, and their subsequent hydrolysis.

At the further entry of metals into the solution (e.g., during the titration with metal salt solutions) and the increase in the Me/ligand ratio, a significant part of WSOS components forming less stable complexes under more acid conditions enters in the reaction with Me. Their content is higher than the content of those formed at the first stage (because of the predominance of WSOS components with average properties). The complexation of Me with these components is also accompanied by a decrease in solution pH by the above reasons. Then, the stability of the formed complexes also decreases. These processes can continue until the protons released in increasing amounts during the complexation of metal ions with WSOSs become capable to actively compete with metal ions for organic acid ligands. Their protonation results in the termination of the complexation of metal ions with WSOS components.

The relationship between the complexation of metal ions with WSOSs and the acid–base properties of the soil liquid phase determines the self-organization (flexibility) of the system of soil solution components, including WSOSs, metals, and protons.

Chapter 8

CHANGES IN THE PROPERTIES OF HUMIC ACIDS UNDER THE EFFECT OF COPPER IONS

The significance of metal load on humic state of soils is provided by numerous mechanisms of their interactions with soil components (Motuzova, 1999; Motuzova and Bezuglova, 2007). But there is little information about the parameters of these processes. One of the first observations of the changes in soil organic matter under the load of heavy metals was noticed (Stanton, 2000), especially when fractionating metal species was used (Minkina et al., 2006, 2008, 2010). These processes that are accompanied by changes of the most important chemical properties of soils have the global significance (Minkina et al., 2009). The first results of this investigation were discussed before briefly (Motuzova et al., 2010). The more detailed analysis of these changes is discussed in this chapter. Informative parameters of humic acids (HAs) include their ^{1}H NMR spectra, molecular weight distribution, and hydrophilic–hydrophobic properties. These properties of HAs were determined in HA preparations from three soil types before and after interaction with copper ions.

CHANGES IN THE ^{1}H NMR SPECTRA OF HAS DURING THE INTERACTION OF HAS FROM THE STUDIED SOILS WITH COPPER IONS

Proton NMR spectra were recorded for the HA fraction >10 kDa before and after its interaction with copper. This HA fraction contained most HA complexes with copper. Under the effect of copper salts, a decrease in the intensity of the signal at σ 0.6 ppm (which is assigned to the protons of the CH_3 group in the γ position, i.e., protons of long aliphatic chains remote from the aromatic ring) was noted in the ^1H NMR spectra of HAs from the three soils (Table 8.1). This effect of copper ions can be related to the detachment of long aliphatic chain fragments from the HA aromatic rings and their subsequent removal from the HA solution during dialysis through a tube with a 10-kDa mesh size membrane.

After the interaction with copper ions, the proportion of protons in aromatic groups in the chernozem HAs increased. This can be due to the relative increase in the content of the corresponding fragments because of the decrease in the content of CH_3 groups by the above reason. The increase in the content of nonsubstituted aromatic structures in HAs at the detachment of long aliphatic chains can also contribute.

Table 8.1. Distribution of protons in different components of HAs from the soils studied (HA fraction >10 kDa) before and after interaction of HAs with copper ions

Sample	Protons of different groups, %				
	aliphatic			methoxyl groups and carbohydrates	aromatic
	γ–CH_n		β–CH_n and α–CH_n	CH_n–O	C_{ar}–H
	0-0.6	0.6-1	1-2.8	2.8-4.0	6.0-9.0
Soddy-podzolic soil HAs	6.0	19.0	24.0	32.0	20.0
Soddy-podzolic soil HAs + Cu	4.0	19.0	23.0	30.0	24.0
Chernozem HAs	5.0	14.0	25.0	40.0	16.0
Chernozem HAs + Cu	2.0	14.0	24.0	40.0	20.0
Sierozem HAs	9.0	18.0	22.0	30.0	21.0
Sierozem HAs + Cu	6.0	21.0	24.0	32.0	17.0

In sierozem HAs, the decrease in the CH_n groups after the interaction with copper ions was accompanied by an increase in the portion of protons of methoxyl groups and carbohydrates, i.e., protons of short and long aliphatic chains at σ 1.0 ppm. This can be related to the high binding capacity of free aliphatic chains and carbohydrates in the presence of copper ions. These processes can be due to the high content of nitrogen-containing components in HAs. The proportion of protons in aromatic rings can relatively decrease in this case.

Changes in the Molecular-Weight Distribution of HA Particles during the Interaction of HAs from the Soil Studied with Copper Ions

The interaction of HAs from three soils with copper ions resulted in an increase in the content of carbon in the HA fraction <10 kDa, which contained free copper ions, copper hydroxy complexes, and copper complexes with low-molecular-weight acids (Table 8.2). This corresponded to changes in the ^{1}H NMR spectra, which indicated the detachment, at the interaction with copper ions, of long aliphatic chains from the structure of HA components >10 kDa capable to pass into the fraction <10 kDa under dialysis conditions. The most significant increase of the carbon content in the HA fractions <10 kDa was observed for HAs of sierozem (by 27%) and chernozem (by 23%).

In the HA fraction >10 kDa, the molecular weight-distribution changed at the interaction with copper ions, which was differently manifested in the HAs of three studied soils (Table 8.3).

In the chernozem HAs, the portion of the high-molecular-weight fraction increased, and the contents of the medium- and low-molecular-weight fractions decreased respectively.

Table 8.2. Content of carbon in the HA fraction <10 kDa before and after interaction with copper ions

HAs of soil	Carbon in the HA fraction <10 kDa, % of C_{HA}		Change in the content of carbon in the HA fraction <10 kDa after interaction with Cu^{2+}, % of $C_{HA\ ini}$
	initial	after interaction with Cu^{2+}	
Soddy-podzolic soil	21.4	24.7	15.4
Chernozem	24.4	30.0	23.0
Sierozem	12.1	15.4	27.3

Table 8.3. Molecular-weight distribution of the HA fraction >10 kDa before and after interaction with copper ions

Sample	Molecular-weight distribution of HAs, %		
	fraction I	fraction II	fraction III
Soddy-podzolic soil HAs	17.5	73.0	9.5
Soddy-podzolic soil HAs + Cu	15.6	79.3	5.1
Chernozem HAs	9.4	85.8	4.8
Chernozem HAs + Cu	12.6	84.0	3.4
Sierozem HAs	27.3	69.3	3.4
Sierozem HAs + Cu	25.1	70.8	4.1

In the HAs of soddy-podzolic soil, the relative content of the medium-molecular-weight fraction increased, and those of the high- and low-molecular-weight fractions decreased respectively. In the sierozem HAs, the portions of the medium- and low-molecular-weight fractions increased, and the portion of the high-molecular-weight fraction decreased.

The increase in the portion of the high-molecular-weight fraction in the chernozem HAs and the portion of the medium-molecular-weight fraction in the HAs of soddy-podzolic soil could be due to the fact that copper ions entering into the interaction with HA molecules can form bridges between HA fragments, which increased their molecular weights.

The decrease in the portion of the low-molecular-weight fraction in the HAs of chernozem and soddy-podzolic soil could be relative because of the decrease in the content of low-molecular-weight particles in the course of the above process, although other mechanisms could also contribute. The decrease in the portions of the high-molecular-weight fraction in the HAs of soddy-podzolic soil and sierozem was also due to the detachment of long aliphatic chains, which in turn resulted in a decrease in the molecular weight of HAs in this fraction.

The possible degradation of complexes of different metals in HAs under the effect of copper ions was indicated by the changes in the concentrations of metals (Fe, Al, Ca, and Mg) in the HA fraction >1 kDa (Table 8.4). At the addition of copper salt to this fraction, its concentration of metals decreased. Copper ions presumably displaced the Fe, Al, Ca, and Mg ions from the initial complexes; these ions passed into the free state and were removed during the dialysis. The copper ions formed more stable complexes with HAs and are retained in the HA fraction >1 kDa.

Table 8.4. Contents of metals (mg/l) in HA and HA + Cu solutions after dialysis through a tube with a 1-kDa mesh size membrane

Solution	Elements (mg/l) in HA and HA + Cu solutions after dialysis through a tube with a 1-kDa mesh size membrane				
	Cu	Fe	Al	Ca	Mg
Soddy-podzolic soil HAs	1.6	10.3	6.2	27.9	6.33
Soddy-podzolic soil HAs + Cu	11.1	5.7	5.7	18.8	3.3
Chernozem HAs	1.7	23.6	9.2	52.5	10.3
Chernozem HAs + Cu	10.4	11.7	9.8	41.6	6.9
Sierozem HAs	2.0	10.4	9.7	31.6	5.9
Sierozem HAs + Cu	10.1	4.5	4.5	24.9	4.4

CHANGES IN HYDROPHILIC–HYDROPHOBIC PROPERTIES OF HAS FROM THE STUDIED SOILS AT THE INTERACTION WITH COPPER IONS

The distribution of amphiphilic HA fractions (>1 kDa) of the studied soils before and after interaction with copper ions is shown in Figure 8.1. The hydrophilic fraction is characterized by the first and second peaks; the other peaks characterize the hydrophobic fraction. A new peak of the hydrophobic fraction appears in all plots. In addition, an increase in the intensity of the last peak of the hydrophobic fraction is observed in the chromatograms of HAs from soddy-podzolic soil and chernozem.

Data in Table 8.5 also point to an increase in the degree of hydrophobicity of the HA fraction >1 kDa at the interaction of HAs with copper ions for all three soils studied. The observed increase in the degree of HA hydrophobicity can be explained by two reasons. One of them is that the interaction of HAs with copper ions involves the functional groups considered to be the main fragments of the hydrophilic fraction.

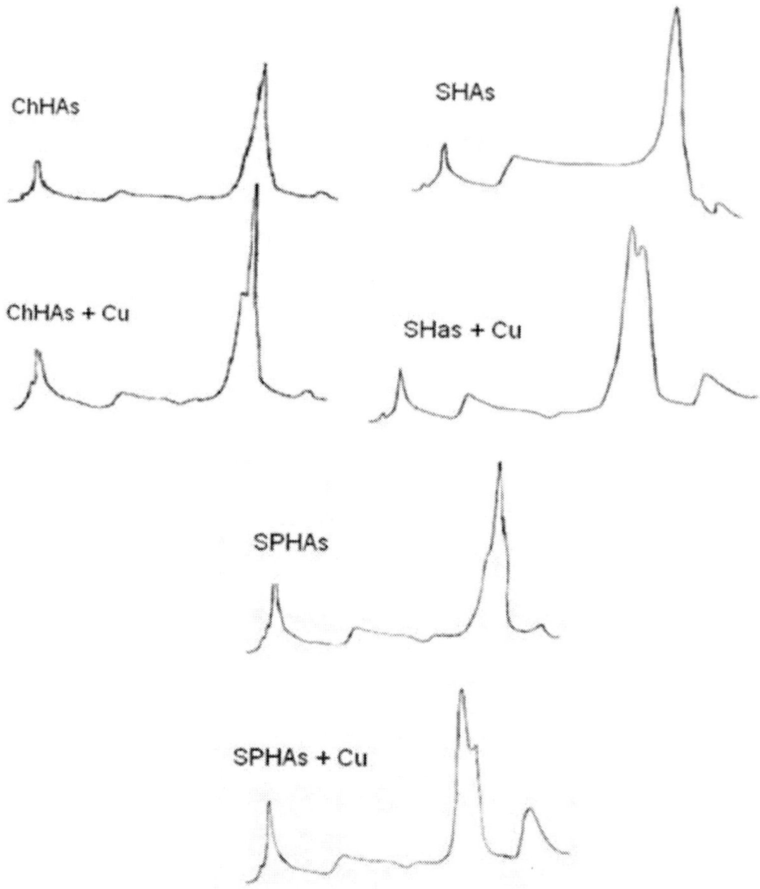

Figure 8.1. Distribution of amphiphilic HA fractions >1 kDa of the studied soils before and after interaction with copper: (ChHAs) chernozem HAs; (SPHAs) soddy-podzolic soil HAs; (SHAs) sierozem HAs.

They form complexes with copper ions, which create metal bridges between HA components, and the content of their free functional groups decreases. This results in a decrease in the content of the hydrophilic fraction.

The other reason for the increase in the degree of HA hydrophilicity can be related to the detachment of long aliphatic chains after the interaction with copper ions (observed in the ^{1}H NMR spectra of HAs) and their subsequent removal from the HA solution by dialysis. These elements are components of the hydrophilic fraction fragments; therefore, the decrease in their content results in a decrease in the content of this fraction.

Table 8.5. Degree of hydrophobicity of the HA fraction >1 kDa before and after interaction with copper ions

HA sample	HA fractions		Hph
	hydrophilic	hydrophobic	
Soddy-podzolic soil HAs	33.63	66.37	1.97
Soddy-podzolic soil HAs + Cu	26.95	73.05	2.71
Chernozem HAs	32. 20	67.80	2.11
Chernozem HAs + Cu	27.89	72.11	2.58
Sierozem HAs	35.74	64.26	1.80
Sierozem HAs + Cu	21.04	78.96	3.75

The most significant increase in the degree of HA hydrophobicity under the effect of copper ions occurred in the sierozem HAs (Hph 3.75) compared to the HAs of chernozem (Hph 2.85) and soddy-podzolic soil (Hph 2.71).

The revealed tendencies in the transformation of humic substances under the effect of interaction with heavy metal ions agree with the concept of Fokin (1975), who revealed exchange between the humic molecules from the solution and those bound to the soil solid phase and called this process "the exchange-sorption renewal of soil humus". In this case, a partial renewal of humus carbon composition occurs in each molecule beginning from its peripheral groups. The author emphasized the self-regulating, stabilizing role of cyclic humus renewal, which could be regulated by humus itself. The external affecting factors could include wetting, the content of fresh organic matter, and other factors primarily affecting microbiological processes.

We think that the addition of extra heavy metal ions, which are active complexing agents, to the system can be one of the external factors affecting the development of humus properties. Their impact on the humus molecule also begins from the most reactive peripheral structures. The substitution of less stable complexes of these structures with typomorphic metal ions by more stable complexes with heavy metal ions (at their addition to the system) can be considered as a type of exchange-sorption processes in the self-regulating humus system. The modification of the humic component in the complexes with metal ions is "the fragmentary renewal of the carbon composition of soil humus".

Chapter 9

Environmental Significance of Humic Substances in Natural and Natural-Anthropogenic Systems

The environmental significance of humic substances in ecosystems is determined by their capacity to affect living organisms both directly and indirectly, through regulation of the state of the environment interacting with the living organisms.

9.1. Effect of Humic Substances on Plants

At the beginning of the 20th century, Bottomley showed that the addition of HSs to nutrient solutions stimulated the growth of plants (Kaschl and Chen, 2005). Later on, this observation was confirmed and supplemented by many authors. It was found that low rates of HSs positively affected the uptake of nitrogen, potassium, phosphorus, and iron by plants (Prozorovskaya, 1936); that the preplant treatment of wheat and corn seeds with humic preparations increased the crop yield by 30–50 kg/ha; and that solanaceous plants were the most responsive to humic fertilizers, and leguminous plants were the least responsive (Khristeva, 1953). It was shown experimentally (Rautan and Schnitzer, 1981) that the topdressing with HAs stimulated the growth of cucumbers and increased the uptake of nutrients by plants. It was found that for corn grown on a nutrient solution, the addition of HAs in concentration of 640 mg/l increased the growth of plants by three times (Tan and Nopamornbodi, 1979); the addition of HAs in concentration of 50 mg/l increased the growth of plants by 36–57% (Vaughan and Malcolm, 1985). The uptake of nutrients by the plants also increased (Tan and Nopamornbodi, 1979; Khaled and Fawy, 2011). However, the relationship between the state of plants and the HS application rate was nonlinear. For plants grown on nutrient solutions, the optimum HS concentration was equal to 50–300 mg/l (van de Venter et al., 1991; Bezuglova, 2001; Kaschl and Chen, 2005). Higher HS concentrations could result in a too strong fixation of nutrients. However, as a rule, such high HS concentrations are not observed in the soil solutions.

Some authors (Pogosyan and Saatov, 1970; Levesque, 1970; Jong and Richmond, 1976; Lee and Bartlett, 1976; Fagbenro and Agbulla, 1993) showed that HSs have a positive effect on the growth of plants on sand or poor soils. They increased the germination rate of seeds, enhanced the uptake of water by seeds, and stimulated the growth of roots and shoots and the elongation of roots (Kaschl and Chen, 2005). The effect of HSs on plants is related to their involvement in physiologically important processes.

9.2. Physiological Significance of Humic Substances

Humic substances are characterized by the high physiological activity. As early as 1951, Khristeva showed that solutions of HSs could serve as sources of nutrients for plants and have an active physiological effect on them. An experimental proof was obtained for the uptake of HSs by plants and the effect of HSs on plants; the effects of different factors were revealed (Pivovarov, 1962; Prat, 1963; Prat and Pospisil, 1959; Khristeva, 1962, 1968; Khristeva et al., 1968).

One of the potential factors of the positive effect of HSs on plants is the capacity of HSs to form complexes with nutrients, which increases their mobility and makes them more available to plants. For example, it was found that the addition of Fe in combination with HSs to a nutrient solution was more efficient than the addition of iron nitrate (Kaschl and Chen, 2005).

The dependence of HS uptake by plants on the molecular mass of HSs and the composition of their functional groups was established. The fractions containing more functional groups and free radicals are more actively absorbed by plants. Thus, the low-molecular-weight humus fractions—FAs—are more actively taken up by plants than HAs (Phuong and Tichy, 1976; Visser, 1986; Aliev, 1987).

Different mechanisms for the effect of absorbed HSs on the plants are considered. The possibility of the hormone-like effect of HAs on the plants was supposed (Muller-Wegener, 1988; Rakovskii and Pigulevskaya, 1978). The effect of HSs on the plasma is possible, and the increase of plasma permeability enhances the uptake of nutrients (Batalkin et al., 1983; Vakhmistrov et al., 1987).

It was found that HSs taken up by plants could affect the essential processes such as respiration and photosynthesis.

More than half a century ago, it was supposed that HSs could act in plant organisms as intermediate catalysts of respiration due to the presence of polyphenols and chinonic groups in the HA structure and their involvement in oxygen exchange (Khristeva, 1951, 1957, 1962). The effect of HS polyphenols on respiration processes is possible, because they can be donors of protons acting as intermediate acceptors of the respiration system (Khristeva, 1968).

In experiments with the application of sodium humates to pea plants, their effect on photosynthesis was revealed, which was due to the impact of HSs on the membrane, phytohormonal, and protein-synthesis systems of plants. The addition of humate accelerated the transfer of electrons in the electron-transport chain of chloroplasts isolated from them, as well as cyclic and noncyclic photophosphorylation. HAs can be directly assimilated by plant cells when reach their nuclei, mitochondria (Fokin, 1975).

Some other mechanisms activating biochemical processes in plants under the impact of HSs are indicated (Visser, 1986):

- effect on ion exchange between protoplasm and soil solution;
- effect on the viscosity and colloidal structure of protoplasm;
- effect on the permeability of cell membranes;
- interaction with enzymes that may be introduced into humus molecules.

Specific mechanisms were revealed for the effect of HA polydispersity on the state of plants. The effect of surface-active properties of HAs on the physiological activity of plants was studied in experiments with cucumbers. When the surface tension of HA solutions decreased (because of an increase in HA concentration) from 72 to 65 d/cm, stimulation of root growth was observed in plants treated with these solutions. However, at the further decrease in surface tension to 61 d/cm, the stimulation of growth was changed by its inhibition. A specific effect of HAs on the cell membranes of plants was supposed (Vakhmistrov, 1987).

The effect of dissolved organic substances (FAs, low-molecular-weight organic acids, vitamins, enzymes, etc.) on the physiological activity of monads in surface water is mainly related to the molecular weights and surfactant properties of dissolved solids. An inhibiting effect of dissolved solids on organisms in undisturbed water bodies was noted in most cases. This phenomenon is explained by the negative effect of the surface-active properties of humus acids (Petersen, 1991). The low-molecular-weight fractions of HSs had a significant negative effect on organisms (Tulonen et al., 1992). However, in the case of contaminated water bodies, dissolved solids mitigate negative changes.

9.3. Effect of HSs on the Agrochemical Properties of Soils

Humus plays the great role in pedogenesis and soil fertility. HSs significantly affect the physical and chemical properties of soils by creating habitat conditions for plants. It is generally known that soils with the high humus content are most fertile (Tyurin, 1965).

HSs largely control the development of physical soil properties and soil structure (Khan, 1969). Both specific and nonspecific HSs are involved in the aggregation of soils, but recently formed HAs and their high-molecular-weight fractions play a special role (Dell`Agnola and Ferrari, 1971). The content and composition of humus are related to the thermal budget of soil, its bulk density, consistence, and the solid phase density (Shein, 2005).

The content of organic matter in the soil affects its color, which in turn affects the thermal budget of the soil. Dark-colored soils are heated more rapidly and retain heat for a longer time. In the zonal soil sequence, the integral reflection of light is mainly determined by the level of humus accumulation. Relationship between the reflective capacity and the humus content is also observed for soils of the same type, although it is less pronounced (Obukhov and Orlov, 1964; Orlov, 1969).

Parameters of different physicochemical properties of soils show significant correlation with the organic matter content.

HSs play a significant role in the development of pH-dependent cation exchange capacity of soils. In the humus horizon, it is mainly determined by the content of organic matter, because the adsorption capacity of HSs significantly exceeds that of mineral soil components (Orlov, 1974).

HSs affect the composition and mobility of adsorbed cations. Organic acids are the main sources of the liquid-phase acidity in the humus horizons of soils. Organic substances affect the redox potential of soils. They are regulators of conditions for the mineral nutrition of plants and ensure the supply of soils with nitrogen, phosphorus, and trace elements (Orlov, 1974).

9.4. Effect of HSs on the State of Agrocenoses

The high positive effect of the application of HSs under experimental conditions gave an impulse to the large-scale production and application of numerous humic fertilizers. Their effect was manifested in two aspects: an increase in the yield of crops was observed, as well as an increase in plant resistance to different unfavorable environmental factors. Abundant evidence of the effect of humic preparations on the state of agrocenoses was accumulated. It was proved that preparations of HAs and potassium and sodium humates increase the field germination of cereal seeds; enhance root formation; increase the absolute grain mass and the content of gluten in the grain (Titova et al., 2007), the productive tillering, the ear length, and the number of grains in the ear (Grebenshchikov, 2003; Kasimova et al., 2003). They also enhance the immune reaction of wheat plants under pathogenesis caused by root rot agents; and increase the resistance of wheat plants to head smut agents (Isaev and Sereda, 2003). The preplant treatment of potato tubers with HS preparations increased the resistance of potato plants to fungal diseases and positively affected the growth and development of potato plants; the treatment of vegetating potato plants with HS preparations inhibited the development of foliar blight and increased the tuber yield (Isaev et al., 2003).

There are different methods for the application of HS preparations: preplant treatment of seeds, foliar dressing of plants, and application into the soil. The efficiency of the methods varies among the crops. For example, the preplant treatment of tubers was found to be most efficient for potato and the foliar dressing of plants, for radish (*Raphanus sativus*) (Puzanov, 2003). Different crops are differently responsive to humic fertilizers: solanaceous plants are most responsive, and legumes are least responsive (Khristeva, 1953). The assessment of the economic efficiency of the preparations showed that the income from the average yield gain exceeded the expenses for the use of preparations by 10–20 times (Puzanov, 2003; Shafronov and Afanas'ev, 2003).

The use of a 0.03% humate solution enriched with microelements at a rate of 500 l/ha for spraying vegetating barley plants increased the crop yield by 430 kg/ha (27%) (Grebenshchikov, 2003). The treatment of seeds with a sodium humate at the HA concentration of 0.75–1.0% (10 l/t) in combination with the spraying of vegetating plants with 0.01% solution of HA increased the wheat yield by 400 kg/ha (Titova et al., 2007). The use of a 0.01% HS solution at an application rate of 0.5 g/m^2 (single root dressing and double spraying of vegetating plants) to cucumbers grown in a greenhouse increased the yield by 2–4 kg/m^2 (10–20% of the control) (Poryvaeva et al., 2003). Application of Fe complexes with HS to growing media effectively prevented white lupin (*Lupinus albus* L.) from Fe chlorosis (de Santiago and Delgado, 2007). Application of 1 g/kg of HA to growing media increased tomato seedling growth and nutrient contents under saline soil conditions (Turkmen et al., 2004).

The preplant treatment of wheat seeds with a sodium humate preparation (0.7 kg/t) increased the field germination of seeds (by 10.7%), the number of fertile tillers, the number of grains in an ear, and the mass of 1000 grains. The grain yield increased by 350 kg/ha, or by 8.3% compared to the control. The spraying of vegetating plants with this preparation (0.2 kg/ha) increased the yield by 4.5% (or 190 kg/ha) as compared with the control. The combined use of two application methods increased the yield by 430 kg/ha (10.1%) (Monakov et al., 2003).

9.4. Significance of Humic Substances for Decreasing the Toxic Effect of Heavy Metals on Living Organisms

The detoxifying effect of metal complexing with dissolved organics (containing a high proportion of FAs) was repeatedly observed in laboratory experiments with water solutions of salts and aquatic test organisms in natural waters with different contents of organic compounds.

Observations of lake waters also provided information on the effect of HAs on the quality of lake water. For example, the photosynthesis of plankton decreased by half, when the concentration of free copper ions in the lake water became $3 \cdot 10^{-7}$ M. However, a different situation was observed in the water of Lake Alpnach, Switzerland, where two third of copper was bound to organic compounds with MM > 1000 Da. Therefore, no decrease in phytoplankton photosynthesis was observed in the lake water at the concentration of copper complexed with organic compounds equal to $1 \cdot 10^{-6}$ M. Dissolved organic substances maintained the concentration of free copper ions in the lake water at a nontoxic level for phytoplankton (Gachter et al., 1978).

A decrease in aluminum toxicity for algae in water under the effect of FAs was noted. FAs increased the tolerance of algae to Al. When the FA concentration in water was 11–22 mg/l, aluminum present at a level of 11–18 μmol/l had no toxic effect on algae (Gardner and Al-Hamdani, 1997).

A series of experiments with aquatic plants was performed. The accumulation of copper and cadmium by *Vallisneria spiralis* L. plants was studied in pot experiments. An HA preparation isolated from coal, which was applied to the soil at a rate 3–8 g/kg, reliably decreased the accumulation of metals by plants: cadmium by 14–50% and copper by 35–77%. In some HA treatments, even an increase in the content of chlorophyll was observed in plants grown under contamination conditions (Wang et al., 2010). A decrease in the accumulation of cadmium and zinc by an aquatic herbaceous plant (*Ceratophyllum denorsum*) under the effect of HAs was described (Bunluesin et al., 2007), as well as a decrease in the uptake of zinc by another aquatic plant (*Lemna minor*), which was accompanied by an increase in the content of chlorophyll in plants and their growth rate (Kruatrachue et al., 2002).

In experiments with water from the Suwannee River (Georgia, the United States), tadpoles of *Xenopus laevis* were used as test objects, and their death was the criterion. Copper was added to the water at concentrations of 0.042–1.18 mg/l. Preparations of FAs (isolated from river water) at concentrations of 0.5 and 5.0 mg/l were used for detoxification. A clear correlation was found between the tadpole death and the activity of copper in the water. At pH 5.5, FAs added at a concentration of 0.5 mg/l had no reliable effect on the activity of copper ions, but their concentration of 5.0 mg/l significantly decreased the activity and toxicity of copper (Buchwalter et al., 1996).

In experiments with seedlings of lettuce (*Lactuca sativa* L.), it was shown that organic compounds capable of complexing with metals could affect plants in two ways. At a copper concentration of 0.16 mmol/l, they could decrease the accumulation of metal by plants, as well as decrease the toxic effect of metal by increasing the resistance of plants to it. The detoxifying effects of different organic substances were compared: HAs, water extracts from soils, complexones (0.16–0.32 mmol/l EDTA and DTPA), and low-molecular-weight organic acids (citric, malic, and oxalic acids). The toxic effect of copper salts was manifested as a decrease in the length of seedling roots and their browning. HAs were found to be the most efficient in decreasing the toxicity. At an HA concentration of 40 mg/l, the toxic effect of copper on plants was completely removed: the length of seedling roots corresponded to that in the control treatment without copper (Inada and Takenaka, 2005).

The results of studying the detoxifying effect of HAs on seedlings of wheat (*Triticum aestivum* L.) germinated in water with a copper concentration of 1 mg/l were reported. The copper toxicity decreased due to the complexation of HAs with copper ions at an HA concentration of 5 mg/l and completely disappeared at HA concentrations of 30 mg/l and higher. At an HA concentration of 30 mg/l, even a stimulating effect of HAs on wheat seedlings was observed compared to the experiments without copper.

Thus, HAs manifested a stimulating effect under metal stress. It is possible that the concentration of free copper ions reached an optimum level for wheat seedlings due to copper complexation with HAs, which explained their stimulating effect (Kulikova, 2008). In this work, modified HA preparations artificially enriched with oxygen-containing functional groups were tested. The detoxifying effect of these preparations was higher than that of the original unmodified HAs, which confirmed the important role of HA oxygen-containing groups in the detoxification of metals. This experiment also showed that the efficiency of HA preparations in the detoxification of metal-contaminated environments could be increased by their artificial enrichment with functional groups.

A significantly smaller number of field studies into the detoxifying effect of HA added to the soil with respect to heavy metals has been performed. Kulikova (2008) conducted a small-plot experiment, in which the effect of HA preparations (applied at 100 kg/ha) on the soil contaminated with copper (750 kg/ha) and their detoxifying effect on cereals (soft wheat) were studied.

At these proportions of copper and HAs in the soil, HAs could not completely eliminate the toxic effect of copper, but they favored a significant increase in the plant biomass in comparison with the control (without HA preparation). On the second year after contamination and HA application, the toxic effect of copper on plants was completely eliminated. The highest detoxifying effect was observed for the preparations artificially enriched with hydroquinone fragments. The protecting effect of HA preparations in soils was lower than in water media because of the higher buffer capacity of soils.

Iron and manganese humates decreased the content of Mo in plants in a pot experiment on Mo-contaminated soils (2.3–11 mg/kg) by 40 and 75%, respectively.

In a field experiment on these soils, manganese humate decreased the content of molybdenum in plants by 25%. Manganese humate was recommended for decreasing the mobility of molybdenum in the soils slightly contaminated with this heavy metal (Neunhauserer et al., 2001).

CONCLUSIONS

Humic acids compose a complex system of heterogeneous polydispersed organic compounds, whose interrelations form the basis for the functioning of this unique natural system. Humic acids are formed under strongly different natural conditions (sources of organic substances, biotic and abiotic conditions of their transformation up to the loss of the original anatomic structure and the formation of new structures) and represent organic substances of specific nature with a wide variety of properties corresponding to the local conditions of their existence. These substances are united by the common elemental composition and the common organization principle of organic components formed by these elements. The almost 70-year-long history of humus study attests to the permanent existence of C, H, O, and N in its composition and the presence of S, P, and metals. The general organization principle of the components formed by them is the presence of an aromatic nucleus and a peripheral carbohydrate–peptide complex.

The analysis of soils from different natural zones with the use of independent methods makes it possible to characterize HAs formed under different pedogenetic conditions. They have some properties characterizing them as a special class of compounds with relatively close proportions of the major elements and the structures formed by them. These structures include an aromatic nucleus and a carbohydrate–peptide periphery with a wide range of functional groups. These components ensure two main types of humus resistance to external impacts. The aromatic component ensures the relative stability of HAs under external impact. The peripheral fragments of humus, i.e., its various functional groups, determine the flexibility of the system of humic substances under different conditions of external impact.

The functional groups of HAs are essential components of their aliphatic part that predetermine the proportions between aromatic and aliphatic fragments in HAs. They significantly affect the hydrophilic–hydrophobic interactions in the soil, which control the adsorption and migration of organic substances and associated elements.

The study of HAs from soils of different natural zones showed that the content of aromatic structures, which mainly determine the manifestation of hydrophobic HA properties, is inferior to that of aliphatic HA components. The hydrophobic properties in HAs of different soils are formed not only due the benzene rings of aromatic fragments but also to the uncharged and nonpolar aliphatic chains, i.e., the amphiphilic aliphatic components of organic substances. The participation of both hydrophobic and hydrophilic components in the formation of humus is a characteristic feature of autochthonous soils.

The functional groups of HAs determine the activity of HAs in their interactions with metals of natural and technogenic origin. Metal ions are active complexing agents. Complexation with metal ions affects not only the state of metals themselves and organic substances but also the ionic composition of soil liquid phase, acid–base conditions, and the ion-exchange and other soil properties. These processes create the chemical status of soils and determine the possibility of the migration and accumulation of organic substances and metals.

The study of water extracts and soil solutions from soils of the taiga zone of the Kola Peninsula subjected to the aerosol emissions from the Severonikel copper–nickel smelter also indicated the effect of these processes. Fulvic acids make up one third of organic substances in water extracts from these soils, with the predominance of nonspecific low-molecular-weight organic acids. Their complexes with copper and nickel ions ensured the predominance of metal particles with negative and zero charges in the liquid phase of these soils. These forms of metal compounds are more capable of migration, and their formation enhances the migration of metals in the soils of the region and results in the local penetration of metals from the soil to the groundwater.

The complexation of metal ions with humic substances can also affect the properties of organic substances themselves. The involvement of functional groups from the peripheral fragments of humus acids in complexation with metals results in the reorganization of the system of humic substances, including the decrease in the content of their aliphatic fragments; the relative increase in the content of the aromatic component; the alteration of the molecular-weight distribution; and the hydrophobization of humic substances. These changes increase the capacity of soils to adsorb metal ions.

The complexation of metal ions with water-soluble organic substances contributes to the formation of the acid–base system in the soil liquid phase and its self-regulation capacity. The complexation of metal ions with organic substances from the liquid phases of soils forms the flexible system of interrelated components: metal–HA ligands–metal–organic complexes–proton. The development of aliphatic chains in HAs and the presence and diversity of functional groups ensure the formation of metal complexes with different binding strength between their components. Their formation is a clearly organized successive involvement of different ligands in complexation. Stable complexes formed at a narrow metal/ligand ratio are displaced by less stable complexes with a wider metal/ligand ratio. The process is accompanied by the regular acidification of solution, which also favors the loosening of the bond between the metal and the ligand. This results in the cessation of complexation and the related acidification of solution.

Thus, the system of humic substances acts as one of the most efficient regulators of essential chemical processes.

REFERENCES

Abd Al-Kareem, Soumi J.G., and Mardoud T. Soil and water resources in the projects of the Euphrates basin. // *The Syrian Economic Journal*. - 1994. - No. 248. – P. 43-70.

Adriano D.C. Trace elements in terrestrial environments. -New York, Berlin, and Heidelberg: Springer-Verlag, 2001.- p. 868.

Afanas'eva E.A. Chernozems of the Russian Upland. Moscow: Nauka, 1966.- p.222.

Alekin O. A. General Hydrochemistry.- Leningrad: Gidrometeoizdat, 1948.- p.206.

Aleksandrova I.V. Procedure of Studying the Qualitative Composition of Organic Substances in Soil Solutions // *Pochvovedenie*.- 1960.-No. 1.-P. 36-44.

Aleksandrova L.N. and Nazarova A.V. Heterogeneity of Humic acids and its genesis. // *Problems of Soil Science: Soviet Soil Scientists to the 11th International Congress in Canada*. Moscow: Nauka, 1978.- P. 12-28.

Aleksandrova L.N. Soil Organic Matter and Its Transformation. – Leningrad: Nauka, 1980.- 287 p.

Aleksandrova L.N., Dorfman E.M., Yurlova O.V. Organo-mineral derivatives of humic substances in the soil. // *Soil Humic Substances: Formation, Composition, Properties, and Significance in Soil Formation and Fertility: Proceedings of Leningrad Agricultural Institute*.- Leningrad,1970.- V. 142.- P. 157-197.

Aleshchukin L.V. Dimensions and ratios of migrated zinc species in soils. // *Zinc and Cadmium in the Environment*.- Moscow: Nauka, 1992.- P.155-180.

Aleshchukin L.V. Geochemistry of copper, nickel, and iron in soils of the polar Murmansk areas. // *Materials on the Geochemistry of Landscapes of the Kola Peninsula.* – 1972.- P.155-180.

Aleshchukin L.V. Transformation and accumulation of metals in tundra soils of Northern Europe. // *Proceedings of the Biogeochemical Laboratory.*- 1990.- V. 21. -P.134-145.

Aliev S.A. Paramagnetism of Soil Organic Matter.- Novosibirsk, 1987.- p.20.

Aristovskaya T. V. The Role of Microorganisms in the Development of the Podzolic Soil Profile. // *Microbiology of Podzolic Soils.* – Leningrad: Nauka, 1965.- P. 141-165.

Arzhanova V. S., Vertel' E. F., and Elpat'evskii P. V. Micronutrients and Soluble Organic Matter in Lysimetric Waters. // *Pochvovedenie.*- 1981.- No. 11.- P.50-60.

Arzhanova V.S. Migration of trace elements in soils from lysimetric data. // *Pochvovedenie.*- 1977. No. 4.- P.71-77.

Bartashevsky Yu.A., Gaydarov O.L., Gordienko S.A. Free radical damping investigation of humic acid with different Mn and Cu content ESR method // *Humus Planta.* - 1971. – No. 5. - P.339-342.

Batalkin G.A., Kochanov M.M., and Makhno M.Yu. Permeability of membranes for some humic substances and their contribution to the physiological activity of Na humate preparation. //*Humic Fertilizers: Theory and Practice of Their Application.* 1983.-V. 8.- P. 117-121.

Beckwith R. S. Metal complexes in soils.// *Aust. J. Agric. Res.* - 1955. - No. 6. - P. 685-698.

Benedetti M.F., Milne C.J., Kinniburgh D.G. et al. Metal ion binding to humic substances: application of the Non-Ideal Competitive Adsorption model // *Environ. Sci. Technol.* – 1995. - V. 29. - P. 446-457.

Beyer L. The chemical composition of soil organic matter in classical humic compound fractions and in bulk samples – a review.// *Z. Pflanzenernähr. Bodenk.*-1996.-No. 159. P. 527–539.

Bezuglova O.S. Humus Status of Soils in Southern Russia. Rostov-on-Don: SKNTs VSh, 2001. - p. 228.

Blaser, P. Sposito G., and Holtzclaw K.M. Composition and acidic group chemistry of an aqueous chestnut leaf litter extract. // *Soil Sci. Am. J.* - 1984. - V. 48. - P.278-283.

Bloom P.R. and McBride M.B. Metal ions binding and exchange with hydrogen ions in acid-washed peat. // *Soil Sci. Soc. Amer. J.* – 1979. - V. 43. - No.4. – P. 687-692.

Brady N.C. and Weil R.R. The Nature and Properties of Soils. - New Jersey: Upper Saddle River, 2002.

Bremner J.M., Heintze S.G., and Mann S.G. Metallo-organic complexes in soil. // *Nature.* – 1946. – No 158. – P. 790-791.

Broadbent F., Ott I. Soil organic matter-metal complexes. I. Fector affecting retention of various cations // Soil Sci. – 1957. – V. 83. – P. 419-427.

Buchwalter D.B., Linder G., Curtis L.R. Modulation of cupric ion activity by pH and fulvic acid as determinants of toxicity in Xenopus laevis embryos and larvae // *Environ. Toxicol. Chem.* – 1996. - V. 15. - P. 568-573.

Bunluesin S., Pokethitiyook P., Lanza G.R., Tyson J., Kruatrachue M., Xing B., Upatham S. Influences of cadmium and zink interaction and humic acid on metal accumulation in Ceratophyllum demersum // *Water, Air and Soil Pollut.* – 2007. - V. 180. - P. 225-235.

Butkevich V. S. and Fedorov M. V. Transformation of Acetic Acid to Succinic and Fumaric Acids in Musor stolonifer Cultures and the Separation and Quantification Methods of These Acids.- Moscow : Akad. Nauk SSSR, 1957.-p. 488.

Candler R., Zech W., and Alt H. G. A Comparison of Water-Soluble Organic Substances in Acid Soils under Beech and Spruce in Ne-Bavaria. // *Z. Pflanzenernaehr. Bodenkunde* - 1989.-V.152.-P.61-65.

Cheam V. and Gamble D.S. Metal-fulvic acid chelation equilibrium in aqueous NaNO3 solution: Hg(II), Cd(II), and Cu(II) fulvate complexes // *Can. J. Soil Sci.* – 1974. - V. 54. - P. 413-417.

Chen, L. A., Dale, R. E., Roth, S., and Brand, L. Nanosecond time dependent fluorescence depolarization of diphenylhexatriene in dimyristoyllecithin vesicles and the determination of 'microviscosity'. //*J. Biol. Chem.* -1977.-V.252.- P.2163-2169.

Cheshire M.V., Berrow M.L., Goodman B.A., and Mundie CM. Metal distribution and nature of some Cu, Mn and V complexes in humic and fulvic acid fractions of soil organic matter. // *Geochim. Cosmochim. Acta.* – 1977. – V. 41. – P. 1131-1138.

Christe I., Knicker H., Kogel-Knabner I., and Kretzschmar R. Chemical Heterogeneity of Humic Substances: Characterization of Size Fractions Obtained by Hollow-Fiber Ultrafiltration. // *Eur. J. Soil Sci.*- 2000.-V.51. No. 4, P.617-625.

Chukov S.N. Structural and Functional Parameters of Soil Organic Matter under Anthropogenic Impact. - St. Petersburg, 2001. – 216 p.

Conte P., Piccolo A. High Pressure Size Exclusion Chromatography (HPSEC) of humic substances. Molecular sizes, analytical parameters and column performance. // *Chemosphere*. -1999.- V.38.- No 3. P.517-528.

Cozzolino A. and Piccolo A. Polymerization of Dissolved Humic Substances Catalyzed by Peroxidase: Effects of pH and Humic Composition.// *Org. Geochem*.- 2002.-No.33.-P. 281-294.

Croue J.-P., Benedetti M.F., Violleau D. and Leenheer J.A. Characterization and copper binding of humic and non-humic organic matter isolated from the South Platte River: Evidence for the presence of nitrogenous binding site. // *Environ. Sci. Technol.* – 2003. – V. 37. – P. 328-336.

De Santiago A. and Delgado A. Effects of humic substances on iron nutrition of lupin // *Biol. Fertil. Soils*. – 2007. – V. 43. – P. 829–836.

Debska B., Drag M., Andersen D. Molecular Size Distribution and Hydrophilic and Hydrophobic Properties of Humic Acids Isolated from Forest Soil. // *Soil and Water Res.* – 2007. – V. 2. – P. 45–53.

Dell`Agnola G., Ferrari G. Molecular sizes and functional groups of humic substances extracted by 0.1 M pyrophosphate from soil aggregates of different stability // *J. Soil Sci.* – 1971. - V. 22. – No. 3.

Demin V.V. The role of humic acids in an irreversible sorption of heavy metals in soil. // *Izv. TSKhA*. – 1994. – No. 2. – P. 79-86.

Dergacheva M.I. Archeological soil science. – Novosibirsk: Nauka, 1997. – 228 p.

Determan H. Gel Chromatography. - New York: Springer, 1968. – p. 220.

Dobrovol'skii V.V. Soil Geography with basic Soil Science. - Moscow: Vlados, 1999.- P. 382.

Dobrovol'skii V.V. and Aleshchukin L.V. Biological cycle and migrated iron and aluminum species. // *Nauchn. Dokl. Vyssh. Shkoly, Biol. Nauki*. - 1977. - No. 9. - P.120-125.

Dobrovol'skii V.V. Heavy metals in the environment. // *Proceedings of the Biogeochemical Laboratory*. - 1990. - V. 21. -P.-88-102.

Dobrovol'skii V.V. Mineralogy and landscape-geochemical characterization of Quaternary deposits on the Kola Peninsula, // *Materials on the Geochemistry of Landscapes of the Kola Peninsula*. -1972.- P.63-68.

Elpat'evskii P. V. and Lutsenko T. N. The Role of Water-Soluble Organic Substances in the Transfer of Technogenic Metals in the Profile of Mountain Burozem. // *Pochvovedenie*.- 1990. -No. 6.-P.30-41.

Essington M.E. *Soil and water chemistry*. An integrative approach. - London. NY. Washington: CRC press. Boca Raton, 2004. - p. 534.

Fagbenro J.A., Agboola A.A. Effect of different levels of humic-acid on the growth and nutrient uptake of teak seedlings // *J. Plant Nutr.* – 1993. - V. 16. – P. 1465-1483.

Fitch A., Stevenson F.J., Chen Y. Complexation of Cu(II) with a soil humic acid: Response characteristics of the Cu(II) ion-selective electrode and ligand concentration effects // *Org. Geochem.* - 1986. - V. 9. - P. 109-116.

Flaig W. Effects of microorganisms on the transformation of lignin ti humic substances // *Geochim. Cosmochim. Acta.* – 1964. – V. 28. – P. 1523.

Flaig W., Beutelspacher H., Rietz E. Chemical Composition and Physical Properties of Humic Substences. // *Soil Components. V. 1. Organic Components.* - Berlin, Heidelberg, New York: Springler-Verlug, 1975. - P. 1111.

Fokin A.D. and Karpukhin A.I. Investigation of the complexes of fulvic acids with Fe(III). // *Izv. TSKhA.* – 1972. – No. 1. – P. 132-137.

Fokin A.D. Studying the transformation, interaction, and transfer of organic substances of iron and phosphorus in podzolic soil. Extended Abstract of Doctoral Dissertation in Biology. - Moscow, 1975.- P. 36.

Forsyth W. Studies on the more soluble complexes of soil organic matter. // *Biochem. J.* - 1947. - V. 47. - No. 2. - P.176-181.

Fox T. R. The Influence of Low-molecular-Weight Organic Acids on Properties and Processes in Forest Soils // *Carbon Forms and Functions in Forest Soils*. - Madison: SSSA, 1993. - P. 43-62.

Fox T.R. and Comerford N.V. Low-molecular-weight organic acids in selected forest soils of the Southeastern USA. // *Soil Sci. Am. J.* - 1990. - V. 54. - No. 4. - P.136-153.

Frund R., Haider K., and Ludemann, H.-D. Impacts of soil managment practices on the organic matter structure investigations by CPMAS 13C NMR-spectroscopy // *Z. Pflanzenernahr. Bodenk.* - 1994. – V. 157. - P. 29-35.

Gachter R., Davis J.S., Mares A. Regulation of copper availability to phytoplankton by macromolecules in lake water // *Environ. Sci. Technol.* – 1978. - V. 12. - P. 1416-1421.

Gamble D.S., Lanford C.H., and Tong J.P.K. The structure and equilibria of a manganese(II) complex of fulvic acid studied by ion exchange and nuclear magnetic resonance. // *Can. J. Chem.* – 1976. – V. 54. - P. 1239-1245.

Ganzhara N.F. and Orlov D.S. Transformation processes of organic matter in soils and its qualitative composition // *The concept of optimization of soil organic matter in agricultural landscapes*. – Moscow: MSKhA, 1993. – P. 18-26.

Ganzhara N.F. Conditions of humus formation and the humus status of zonal soil types // *Izv. Timiryaz. S-Kh Akad.*- 1986.- No. 5.- P. 84-89.

Gardner J.L., Al-Hamdani S.H. Interactive effects of aluminium and humic sustances on salvinia // *J. Aquat. Plant Manage.* – 1997. - V. 35. - P. 30-34.

Gel'tser F.Yu. The significance of microorganisms in the formation of humus and strength of soil structure. – Moscow: Sel'khozgiz, 1940. – 190 p.

Ghosh K., Schnitzer M. Macromolecular structures of humic substances. // *Soil Science*.-1980.- V. 129.-P. 266-276.

Glazovskaya M.A. Soils of the World. V. 2. - Moscow: MGU, 1973. - p.425.

Gran G. Determination of the equivalent point in potentiometic titration. // *Acta Chem. Scand.* – 1952. – V. 4. - P. 559-577.

Grebenshchikov V.Yu. Use of humic preparations Gumat-80 and Gumat-7 in the growing of barley. // *Proceedings of the II International Conference "Humic Substances in the Biosphere*". - Moscow, 2003. - P. 102-104.

Gregor T.E., Powell H.K.J., and Town R.M. Evidence for aliphatic mixed mode coordination in copper(II)-fulvic acid complexes. // *Journal of Soil Sci.* – 1989. – V. 40. - P. 661-673.

Grishina L.A. Humus Formation and Humus Status of Soils. Moscow, 1986.- p. 242.

Guggenberger G., Kogel-Knabner I., Ufumaier L., *et al.* Gel permeation chromatography of water-soluble organic matter with deionized water: II. Spectroscopic and chemical characterization of fractions obtained from an aqueous litter extract. //*Sci. Total Environ.*- 1989.-V. 81-82, P.447-457.

Hayes M. H. B. and Clapp C. E. Humic Substances: Considerations of Composition, Aspects of Structure, and Environmental Influence // *Soil Sci.* - 2001. - V. 166. – No 11. - P. 723-737.

Heintze S. G. and Mann P. J. G. Soluble complexes of manganic manganese. // *J. Agric. Sci.* - 1947. - No. 37. - P. 23-26.

Hodson J.F., Geering H.R., and Norvell W.F. Micronutrient cation complexes in soil solutions: partition between complexed and uncomplexed forms by solvent extraction // *Soil Sci. Am. Proc.* - 1965. - V. 29. - No. 3. - P. 33-45.

Huang P. M. and Violante A. Influence of Organic Acids on Crystallization and Surface Properties of Precipitation Products of Aluminum. // *Interaction of Soil Minerals with Natural Organics and Microbe.* / Ed. by P. M. Huang and M. Schnitzer.- Madison: SSSA, 1986.- P. 159-221.

Ilaiwi M. Contribution to the knowledge of the soils of Syria. // *PhD Thesis.* – Belgium: University of Ghent, 1983.

Inada S. and Takenaka C. Effects of dissolved organic matter on toxicity and bioavailability of copper for lettuce sprouts // *Environ. Int.* – 2005. - V. 31. - P. 603-608.

Isaev R.F. and Sereda N.A. Effect of different Gumi forms on the phytosanitary state of spring wheat plantations. // *Proceedings of the II International Conference "Humic Substances in the Biosphere"*. - Moscow, 2003. - P. 108-109.

Isaev R.F., Yamaleev A.M., Balakhontsev E.N., and Iskhakov F.F. High-performance immune stimulator. // *Proceedings of the II International Conference "Humic Substances in the Biosphere"*. - Moscow, 2003. - P. 109-110.

Ivanova S.E., Sokolova T.A. Methods for the presentation of data, reproducibility, and information value of the continuous potentiometric titration of soil suspensions and soil extracts // *Vest. Mosk. Univ., Ser. 17: Pochvoved.* – 1998. - N. 4. - P. 16-22.

Jong L.S. and Richmond B.I. Stimulation of plant growth by humic substances // *Soil Sci. Soc. Am. J.* – 1976. - V. 40. - P. 876-879.

Kachl A., Chen Y. Interactions of humic substances with trace metals and their stimulatory effects on plant growth // *Use of humic substances to remediate polluted environments: from theory to practice.* / Ed. by Perminova I.V. et al. - NATO Science Series: IV: Earth and Environmental Sciences. Vol. 52. – Dordrecht: Springer, 2005. – P. 83-113.

Kaemmerer M., Guiresse M., Revel J.C. et al. Conductometric behaviour of humic acids with Cu(II) ions. // *Analusis.* - 1999. - V. 27. - No 5. - P. 421-423.

Kalabin G.A., Kanitskaya L.V., and Kushnarev D.V. Quantitative NMR Spectroscopy of Natural Organic Raw Materials and Their Derivatives. - Moscow: Khimiya, 2000. - p. 498.

Karavanova E.I., Belyanina L.A., and Stepanov A.A. Water-soluble organic matter and soil solution acidity in the main soil types of the central forest state biosphere reserve // *Eurasian Soil Science*. – 2007. – V. 40. – P. 493-504.

Kasimova L.V., Perchenko N.A., Glagol'ev V.P., Titova E.V., Kravets A.V. The influence of humate preparations on wheat yield // *Proceedings of the II International Conference "Humic Substances in the Biosphere"*. - Moscow, 2003. – P. 111-112.

Kaurichev I.S. and Nozdrunova E.M. Role of water-soluble organic matter from plant residues in the formation of mobile iron-organic complexes. // *Pochvovedenie*. - 1961. - No. 10. – P.10-18.

Kaurichev I.S., Ivanova T.N., and Nozdrunova E.M. Low-Molecular Organic Acids in Water-Soluble Organic Matter of Soils. // *Pochvovedenie*. - 1963 - No. 3. - P. 27-35.

Kaurichev I.S., Karpukhin A.I., and Yashin I.M. System study of coordination compounds of soil organic matter. // *Abstracts of the IV Congress of the Dokuchaev Soil Science Society*. - Novosibirsk. - 2004. - V. 2. - P.45-47.

Kaurichev I.S., Nozdrunova E.M., and Tsurupa I.G. Chelatic Fe-organic compounds in soils. // Reports of VIII International Congress of Soil Scientists. – Moscow: Nauka, 1964. – P. 158-166.

Kaurichev I.S., Yashin I. M., Karpukhin A.I., and Platonov I.G. Mobilization and Transformation of Water-Soluble Organic Substances in Podzolic Soils. // *Izv. Timiryazevsk. S-Kh. Akad*.- 1991.- No. 3.- P.71-84.

Kaurichev I.S., Yashin I.M., and Chernikov V.A. Theory and Practice of Sorption Lysimeters in Ecological Studies. – Moscow: MSKhA, 1996. - p. 144.

Khaled H. and Fawy H.A. Effect of different levels of humic acids on the nutrient content, plant growth, and soil properties under conditions of salinity. // *Soil and Water Res*. – 2011. - V. 6. – P. 21–29.

Khan D.V. Organo-Mineral Compounds and Soil Structure. - Moscow: Nauka, 1969. - 142 p.

Khan D.V. Procedure for studying the composition of water-soluble substances from plant residues. // *Tr. Pochv. Inst. im. V.V. Dokuchaeva*. - 1951. - V. 38. - P.99-107.

Khmel'nitskii R.A. Current Methods of Studying Agronomic Objects. - Moscow: *Vyssh. Shkola*, 1981. - p.280.

Khristeva L.A. Function of humic acids in the metabolism of higher plants. // *Humic Fertilizers: Theory and Practice of Their Application*. – Dnepropetrovsk, 1962. - V. 2. - P. 123-129.

Khristeva L.A. Physiological function of humic acids in the metabolism of higher plants. // *Humic Fertilizers: Theory and Practice of Their Application*. – Dnepropetrovsk, 1957. - P.

Khristeva L.A. Role of humic acid in plant nutrition and humic fertilizers. // *Tr. Pochv. Inst. Im. V.V. Dokuchaeva*. - 1951. - V. 38. - P. 108-184.

Khristeva L.A. Role of humic acids and other organic compounds in the nutrition of hicher plants // *Pochvovedenie*. - 1953. - No.10. - P. 24-29.

Khristeva L.A. The nature of the effect of physiologically active humic acids and other plant growth stimulators. // *Humic Fertilizers: Theory and Practice of Their Application.* – Dnepropetrovsk, 1968. - V. 3. - P. 13-27.

Khristeva L.A., Starostin A.N., and Ulitina V.P. The nature of the physiological activity of humic acids. // *Humic Fertilizers: Theory and Practice of Their Application.* – Dnepropetrovsk, 1968. - V. 3. - P. 136-142.

Kolosov I.V. Humic acids as multi-centered high-molecular-weight complexing agents // *Pochvovedenie.* – 1982. - N. 4. - P. 42-48.

Kononova M.M. Soil Humus and Current Problems in Its Study. - Moscow, 1951.-p.385.

Kononova M.M. Soil Organic Matter, Its Nature, Properties, and Methods of Study.- Moscow: AN SSSR, 1963. - p. 314.

Kononova M.M. Studying the recent formation of humic substances // *Pochvovedenie*. - 1944. - No. 10. - P. 456-470.

Kovda V.A. Basic Soil Science.- Moscow: Nauka, 1973. - V.2. - P. 467.

Krainov S.R. Geochemistry of Rare Elements in Groundwaters. - Moscow: Nauka, 1973. - p. 296.

Kravkov S.P. Biochemistry and agrochemistry of soil processes. – Leningrad: Nauka, 1978. – 291 p.

Kretovich V. L. Basic Biochemistry of Plants. – Moscow: Vysshaya Shkola, 1971. - p.465.

Kruatrachue M., Jarupan W., Chitramwong Y.P., Pokethitiyook P., Upatham E.S. Parkpoomkamol K. Combined effects of lead and humic acid on growth and lead uptake of duckweed, Lemna minor // *Bull. Environ. Contam. Toxicol.* – 2002. - V. 69. - P. 655-661.

Kulikova N.A. Protecting effect of humic substances for plants in water and soil environments under abiotic stresses. Doctoral Dissertation. - Moscow, 2008. - p.302.

Ladonin D.V., Reshetnikov S.I., Sadovnikova L.K., et at. Copper ion activity in contaminated and background soils under model experimental conditions. // *Pochvovedenie.*-1994.- No. 8.- P.46-52.

Larina N.K. and Kasatochkin V.I. Infrared spectra of soil humic acids and the table of wave numbers. // *Organic Matter of Virgin and Developed Soils: Experimental Data and Methods of Study.* - Moscow: Nauka, 1972. - p. 245-256.

Lee Y.S., Bartlett R.J. Stimulation of plant growth by humic substances // *Soil Sci. Soc. Am. J.* – 1976. - V. 40. - P. 876-879.

Levesque M. Fulvic acid and fulvometallic complexes in mineral nutrition of plants // *Can. J. Soil Sci.* – 1970. - V. 50. - P. 385-390.

Lindsay W.L. Chemical equilibria in soils. - N.Y.: Wiley-Interscience, 1979. - p.449.

Linnik P.N. and Vasilchuk T.A. Role of humic substances in the complexation and detoxification of heavy metals: case study of the Dnieper reservoirs // *Use of humic substances to remediate polluted environments: from theory to practice.* / Ed. by Perminova I.V. et al. - NATO Science Series: IV: Earth and Environmental Sciences. Vol. 52. – Dordrecht: Springer, 2005. – P. 135-154.

Linnik P.N., Nabivanets B.I. Metal migration forms in fresh surface water. - Leningrad: Gidrometeoizdat, 1986. – 270 p.

Lobartini, J.C., Orioli G.A., and Tan K.H. Characteristics of soil humic acid fractions separated by ultrafiltration. // *Comm. Soil Sci. Plant Anal.* - 1997. – V. 28. – P. 787-796.

Lobartini, J.C., Tan K.H., Rema J.A., Gingle A.R., Pape C. and Himmelsbach D.S. The geochemical nature and agricultural importance of commercial humic matter. // *Sci. Tot. Environ.* - 1992. – V. 113. – P. 15.

Logan E.M., Pulford I.D., Cook G.T., and Mackenzie A.B. Complexation of Cu^{2+} and Pb^{2+} by peat and humic acid. // *European J. Soil Sci.* – 1997. – V. 48. - P. 685-696.

Lukina N. V. and Nikonov V. V. Biogeochemical Cycles in Northern Forests under Aerotechnogenie Contamination. - Apatity: Ross. Akad. Nauk, 1996. - V. 1 - p.213.

Lukina N. V., Gorbacheva T. T., and Nikonov V. V. Spatial Variability of Soil Acidity in Al-Fe-humus Podzols of Northern Taiga. // *Eur. Soil Sci.*-2002.-V.35.-No 2.-P.144-155.

Lundstrom U.S., van Breemen N., Bain D.C. et al. Advances in understanding the podzolization process resulting from a multidisciplinary study of thee coniferous forest soils in the Nordic Countries // *Geoderma.* – 2000. - V. 94. – P. 335-353.

Lur'e Yu.Yu. Reference Book on Analytical Chemistry.- Moscow: Khimia, 1989.- 448 p.

Lykov A.M. Transformation of organic matter and nitrogen in soddy-podzolic soil in a long-term experiment of the Timiryazev Agricultural Academy // *Pochvovedenie*.- 1973.- No. 11.- P. 53-61.

Lysimeters in Soil Studies./ Ed. by L. L. Shishov .- Moscow: *Pochv. Inst. im. V.V. Dokuchaeva*, 1998.- p. 246.

MacCarthy P. The Principles of Humic Substances. // *Soil Sci.*- 2001.- V.166.- No 11.-P.738-751.

Makharadze G.A., Varshal G.M., and Supatoshvili G.D. Complexation of copper ions with fulvic acids isolated from natural waters. // *Chemical Analysis of Marine Sediments.*- Moscow: Nauka, 1988. P.61-68.

Manskaya S.M., Drozdova T.V., and Emel'yanova M.P. Immobilization of copper by different natural organic compounds. // *Pochvovedenie.*- 1958. - No. 6.- P.41-48.

Martell A.E. and Calvin M. Chemistry of the Metal Chelate Compounds. - New York: Prentice-Hall, 1952. -P.352.

Martin J. P., Haider K. Microbial Activity in Relation to Soil Humus Formation. // *Soil Science.*-1971.- V. 111.- No 1.-P. 54-63.

McBride M.B. Transition metal bonding in humic acid. An ESR study. // *Soil Sci.* – 1978. – V. 126. - P. 200-209.

Milanovskii E.Yu. Amphiphilous Components of Soil Humic Substances. // *Eur. Soil Sci.*- 2000.-V.33.-No.6. –P.617-625.

Milanovskii E.Yu. Soil Humus Substances as Natural Hydrophobic-Hydrophilic Compounds.- Moscow: GEOS, 2009.- p.185.

Milanovskii E.Yu., Shein E.V., and Stepanov A.A. Lyophilic-Lyophobic Properties of Organic Matter and the Soil Structure // *Pochvovedenie.*- 1993.- No. 6.-P.122-126.

Min'ko O.I., Izmailova N.G., Milanovskii E.Yu., and Orlov D.S. Gel-chromatographic analysis of water-soluble organic matter // *Nauchn. Dokl. Vysch. Shk., Biol. Nauki.*- 1991.-V. 4.- P. 117-125.

Mina V.N. Effect of Precipitation Flowing down Tree Stems on the Soil.- *Pochvovedenie*, -1967.-No. 12.-P.44-52.

Minkina T.M., Motuzova G.V., Nazarenko O.G. and Mandzhieva S.S. Heavy Metal Compounds in Soil. Transformation upon Pollution and Ecological Significance. - New York : Nova Science Publishers, Inc, 2009.- P.183.

Minkina T.M., Motuzova G.V., Nazarenko O.G.,Kryshchenko V.S. and Mandzhieva S.S. Combined approach for fractioning metal compounds in Soils. // *Eur.Soil Sci.*-2008 .-V.41.-No 11.-P. 1170-1178.

Minkina T. M., Motuzova G. V. and Nazarenko O. G. Interaction of heavy metals with the organic matter of an ordinary chernozem // Eur. Soil Sci.- 2006.- No.7.- P. 702-710.

Minkina T.M., Motuzova G.V., Nazarenko O. G. and Mandzhieva S. S.Heavy Metal Compounds with Organic Substance and Methods of their Definition.// *Materials of 15th Meeting of the IHSS "Humic Substances and the Maintenance of Ecosystem Services". Tenerife, Spain,* 2010.- V.3.

Mirchink T.G. Soil mycology. – Moscow: MGU, 1976. – 204 p.

Monakov S.B., Subkhankulov A.A., and Krylova L.G. Sodium humate as a stimulator of wheat yield. // *Proceedings of the II International Conference "Humic Substances in the Biosphere,"* - Moscow, 2003. - P. 118-119.

Moore J.and Ramamoorthy S. Heavy Metals in Natural Waters.- Moscow: Mir. 1987.p.285.

Motuzova G.V. Microelements in Soils: System Organization, Ecological Significance and Monitoring.- Moscow: URSS,1999.- p.186.

Motuzova G.V., Bezuglova O.S. Ecological Monitoring of Soils.- Moscow: Gaudeamus, 2007.- p.237.

Motuzova G.V., Dergam H., Stepanov A.A The Interaction of Cu^{2+} With Humic Acids of Different soil. // *Materials of 15th Meeting of the IHSS "Humic Substances and the Maintenance of Ecosystem Services". Tenerife, Spain*.- 2010.-V.3.A

Motuzova G.V., Lukina N.V., Nikonov V.V., et al. Effect of natural and technogenic factors on soils, soil waters, and groundwaters of the Kola Peninsula. // *Vodn. Res.* -2004.-No. 3, P.3-5.

Motuzova G.V., Zorina A.V., Stepanov A.A. Water-soluble organic substances in soil water of taiga biogeocenoses of the Kola Peninsula // *Humic Substances in the Biosphere: Proceedings of the II International Conference*. - Moscow. - 2003. - P. 57-58.

Muller-Wegener U. Interaction of humic substances with biota. // Humic substances and their role in the environment. / Edited by Frimmel, F.H. and Christman, R.F. -John Wiley and Sons Limited, 1988.- P. 179-192.

Murphy E.M., Zachara J.M., Smith S.C., and Phillips J.L. The sorption of humic acids to mineral surfaces and their role in contaminant binding // *Sci. Total Environ.* – 1992. – V. 117/118. – P. 413-423.

Neunhauserer C., Berreck M., and Insam H. Remediation of soils contaminated with molybdenum using soil amendments and phytoremediation // Water, Air and Soil Pollut. – 2001. – V. 128. – P. 85-96.

Nikonov V.V. and Lukina N.V. Biogeochemical Functions of Forests at Their Northern Limit.- Apatity, 1994.- p.315.

Nikonov V.V. and Pereverzev V.N. Pedogenesis in the Subarctic Kola Regions.- Leningrad, 1989.-p.168.

Obukhov A.I. and Orlov D.S. Spectral reflectance of the main soil types and the use of diffuse reflectance in soil studies // Pochvovedenie. – 1964. No. 2. – P.

Orlov D.S. and Grishina L.A. Laboratory Manual for Humus Chemistry. – Moscow: MGU, 1981.p. 272

Orlov D.S. and Milanovskii E.Yu. Gel Chromatography in Soil Science: Potentials and Limitations. //*Current Physical and Chemical Methods of Soil Examination.*- Moscow, 1987, P. 94-118.

Orlov D.S. and Osipova N.N. Infrared Spectra of Soils and Soil Components. – Moscow: MGU, 1988. – p. 86.

Orlov D.S. and Sadovnikova L.K. Soil organic matter and protective functions of humic substances in the biosphere // *Use of humic substances to remediate polluted environments: from theory to practice.* / Ed. by Perminova I.V. et al. - NATO Science Series: IV: Earth and Environmental Sciences. Vol. 52. – Dordrecht: Springer, 2005. – P. 37-52.

Orlov D.S. Properties and functions of humic substances. // *Humic Substances in the Biosphere.* - Moscow: Nauka, 1993. - P. 16-27.

Orlov D.S. Soil Chemistry. – Moscow: MGU, 1992. – 400 p.

Orlov D.S. Soil Humus Acids and the General Theory of Humification. - Moscow: MGU, 1990. - P. 324.

Orlov D.S. Soil Humus Acids. – Moscow: MGU, 1974. – 331 p.

Orlov D.S. Variation of organic matter content and redox potential in the plow horizon of soddy-podzolic soil. // Nauchn. Dokl. Vysch. Shk., Biol. Nauki. – 1969. - No. 3.

Orlov D.S., Biryukova O.N., and Sukhanova N.I. Organic Matter in Soils of Russian Federation. – Moscow: Nauka, 1996. – 253 p.

Orlov D.S., Min'ko O.I., and Demin V.V. On the order and mechanisms of formation of metal-humus complexes // Pochvovedenie. – 1988. – No. 9. – P. 43-52.

Osterberg R., Wei S., Shirshova L. Inert copper ion complexes formed by humic acids // *Acta Chem. Scand.* – 1999. - V. 52. - P. 172-180.

Panikov N. S., Sadovnikova L. K., and Fridland E. V. Nonspecific Compounds of Soil Humus.- Moscow, 1984.-p.144.

Perminova I. V., Yudov M. V., Kovalevsky D. V., et al. ^{12}C and ^{1}H NMR Spectroscopic Studies of the 0.2 M HCI Hydrolyzed Humic and Fulvic Acid. // *Proceedings of the 2nd European Symposium: NMR in Soil Science, Freising, Germany, 2000.*-Freising.

Perminova I.V. Analysis, classification, and prediction of humic acid properties. / Doctoral Dissertation in Chemistry. - Moscow, 2000. - 359 p.

Perminova I.V. and Hatfield K. Remediation chemistry of humic substances: theory and implications for technology // Use of humic substances to remediate polluted environments: from theory to practice. – Ed. by

Perminova I.V. et al. - NATO Science Series: IV: Earth and Environmental Sciences. Vol. 52. – Dordrecht: Springer, 2005. – P. 3-36.

Petersen, R.C., Jr. The contradictory biological behavior of humic subsstances in the aquatic environment. // Humic substances in the aquatic and terrestrial environment.- Berlin Heidelberg: Springer-verlag, 1991.- P. 369-389.

Phuong H.K. and Tichy V. Activity of humus acids from peat as studied by means of some growth regulator bioassays // Biol. Plant. – 1976. – V. 18. – P. 195-199.

Piccolo A. and Stevenson F.J. Infrared spectra of Cu, Pb and Ca complexes of soil humic substances. // *Geoderma*. – 1982. - V. 27. - No.3. - P. 195-208.

Pivovarov L.R. The nature of the physiological activity of humic acids in relation to their structure. // *Humic Fertilizers: Theory and Practice of Their Application*. – Dnepropetrovsk, 1962. - Vol. 2. - P. 101-121.

Pogosyan K. and Saatov R. Some results of testing carbon-humic fertilizers on eroded sierozems under cotton plantations. // *Theoretical Principles of the Effect of Physiologically Active Substances and the Efficiency of Their Fertilizers*. - Dnepropetrovsk, 1970. - P. 28-37.

Ponomareva V.V. Theory of Podzol Formation. - Moscow: AN SSSR, 1964. - 379 p.

Popov A. I. Humic Substances: Properties, Structure, and Formation. -St. Petersburg: StPU, 2004.-p.246.

Poryvaeva O.V., Kasimova L.V., and Kravets A.V. Use of humic stimulators on the basis of peat in protected horticulture. // *Proceedings of the II International Conference "Humic Substances in the Biosphere"*. - Moscow, 2003. – P. 122.

Prat S. and Pospisil F. Humic acids with C14 // Biol. Plant. – 1959. – V. 1. – P. 71-80.

Prat S. Permeability of plant tissues to humic acids // Biol. Plant. – 1963. – V. 5. – P. 279-283.

Preston C.M. and Rusk A.C.M. A bibliography of NMR applications for forestry research / Information Report BC-X-322. – Victoria: Forestry Canada, Pacific Forestry Centre, 1990. - 49 p.

Prozorovskaya A.A. Effect of humic acid and its derivatives on the intake of phosphorus, potassium, and iron in plants // *Sb. NIIU-IF*. – 1936. - V. 127. - P. 133-141.

Puzanov V.I. Practice of the regular application of humates in Irkutsk oblast. // *Proceedings of the II International Conference "Humic Substances in the Biosphere"*. - Moscow, 2003. – P. 124.

Rakovskii V.E, and Pigulevskaya L.V. Chemistry and Genesis of Peat. - Moscow: Nedra, 1978. - 231 p.

Rauthan B.S., Schnitzer M. Effects of a soil fulvic acid on the growth and nutrient content of cucumber (Cucumis sativus) plants // *Plant and Soil.* – 1981. - V. 63. P. 491-495.

Robertson A.P., Leckie J.O. Acid/base, copper binding, and Cu^{2+}/H^{+} exchange properties of a soil humic acid, an experimental and modeling study. // *Environ. Sci. Technol.*- 1999.- V. 33.- No. 5.-P.2519-2530.

Rode A.A. Some data on the physicochemical properties of water-soluble organic substances in forest litters // *Pochvovedenie.*- 1941.-No. 3.- P.103-126.

Rodin L.E. and Bazilevich N.I. Dynamics of Organic Matter and the Biological Cycle in the Main Types of Vegetation. - Moscow: Nauka, 1965. - 253 p.

Saar R.A., Weber J.H. Complexation of cadmium(II) with water- and soil-derived fulvic acids: effect of pH and fulvic acid concentration // *Can. J. Chem.* – 1979. - V. 57. – P. 1263-1268.

Saiz-Jimenez C. The chemical structure of humic substances: resent advances. // Humic substances in terrestrial ecosystems. /Ed. by Piccolo A. – Amsterdam: Elsevier, 1996. – P. 1-44.

Samoilova E.M., Sizov A.P., and Yakovchenko V.N. Soil organic matter in the chernozem zone. – Kiev: Naukova Dumka, 1990. – 120 p.

Santore R.C., Di Toro D.M., Paquin P.R. et al. Boitic ligand model of the acute toxicity of metals. 2. Application to acute copper toxicity in freshwater fish and Daphnia // *Environ. Sci. Tech.* – 2001. – V. 10. – P. 2397-2402.

Schnitzer M. Humic substanses: chemistry and reactions. // Soil Organic Matter. / By M. Schnitzer and S. U. Khan.- Amsterdam: Elsevier, 1978.-P.1-64.

Schnitzer M. Recent advances in humic acid research.// Proc, Int. Peat Symp. Bemidji. Minn. Oct. 21-23, 1981. - Bemidji, Minn, 1981. - P. 17-44.

Schnitzer M.and Kerndorf H. Reactions of fulvic acid with metal ions. //*Water Air Soil Pollut*- 1981.-V. 15.- P.68-79.

Schulten R. Interactions of dissolved organic matter with xenobiotic compounds: Molecular modeling in water. //*Environ. Toxicol. Chem.*- 1999.-V. 18.-P. 1641-1655.

Senesi N. Molecular and quantitative aspects of the chemistry of fulvic acid and its interactions with metal ions and organic ions and organic chemicals. // Anal. Chim. Acta. – 1990. – V. 232. – P. 77-106.

Senesi N., Loffredo E. Metal ion complexation by soil humic substances. // Chemical Processes in Soils. / Ed. by Tabatabai M. A. and Sparks D. L. – Madison: Soil Science Society of America, 2005. – P. 563-617.

Shafronov O.D. and Afanas'ev V.V. Efficiency of Gumi for spring wheat in a competitive production test. // *Proceedings of the II International Conference "Humic Substances in the Biosphere"*. - Moscow, 2003. – P. 129.

Shein E.V. and Milanovskii E.Yu. The Role of Organic Matter in the Formation and Stability of Soil Aggregates. // *Eur. Soil Sci.*- 2003.-V.36.- No.1.-P.51-58.

Shein E.V. The course of soil physics. – Moscow: MGU. – 2005. – 432 p.

Shido H.and Kuwatsuka S. Behavior of Phenolic Substances in the Decaying Process of Plants.//Soil Sci. and Plant Nutr. 1977.-V.23.- No 3.- P.88-96.

Skrynnikova I.N. Soil solutions of the southern forest zone and their role in recent pedogenesis processes // Recent Soil Processes in the Forest Zone of the European Soviet Union.- Moscow : Akad. Nauk SSSR, 1959.- P. 136-158.

Sokolova T.A., Tolpeshta I.I., Trofimov S.Ya. Soil acidity and the acid-base buffer capacity of soils. Aluminum Compounds in the Soil Solid Phase and Soil Solution. - Tula, 2007. - 95 p.

Soldatenkov S. V. Organic Acid Exchange in Plants. // *The 30th Timiryazev Lectures.*- Leningrad: Nauka, 1971.-p.148.

Stanton S. Direct and indirect effects of Organic Matter on Metal immobilization in soil. // *Soil-Mineral- Organic Matter-Microorganism Interaction and Ecosystem Health. Third International Symposium of the Working Group MO.* Naples-Capri, Italy.- 2000.- P.41

Stepanov A. A. Influence of manure drains of the pork complex on humus state of meadow gley soil and property of humic substances. Extended Abstract of Candidate's Dissertation in Biology.-Moscow, 1995.- p.24.

Stepanov A.A. and Milanovskii E. Yu. Hydrophobic Properties of Organic Matter in Some Soils of the Ubsu-Nur Basin. // *Proceedings of the Multilateral Workshop of COMECON Countries: the Ubsu-Nur Soviet-Mongolian Experiment.*- Pushchino, 1989. P. 53-54.

Stepanov A.A., Zharkova L.V., and Stepanova E.A. Application of ^1H NMR Spectroscopy for the Characterization of Humic Substances. // *Eur. Soil Sci.*-1997.-V.30.-No2.- P.142-145.

Stevenson F. J. Humus Chemistry.- New York: Wiley, 1982.- p.365.

Stevenson F.J. Humus Chemistry: Genesis, Composition, Reactions. 2nd edn.- New York: John Wiley and Sons Inc., 1994.- P. 496.

Stevenson F.J. and Fitch F. Reactions with organic matter. // *Copper in Soils and Plants*.- New York: Academic press, 1981.-p.69.

Stevenson F.J. Stability constants of Cu^{2+} and Cd^{2+} with humic acids.// *Soil Sci Amer J*.-1976.-V. 40.- No. 5.-P.665-672.

Supatashvili G.D. and Makharadze G.A. Species of elements in natural waters and their dependence on ionic potentials. // *Chemical Analysis of Marine Sediments*.- Moscow: Nauka, 1988. P.52-61

Takamatsu T. and Yoshido T. Determination of stability constants of metal-humic acid complexes by potentiometric titration and ion-selective electrodes // *Soil Sci*. – 1978. – V. 125. - No. 6. - P. 377-386.

Tan K.H., Nopamornbodi V. Effect of different levels of humic acids on nutrient content and growth of corn (Zea mays L.) // *Plant and Soil*. – 1979. - V. 51. - P. 283-287.

Tani M., Higashi T., and Nagatsuka S. Dynamics of Low-molecular-Weighl Aliphatic Carboxylic Acids (LACAs) in Forest Soils: I. Amount and Composition of LACAs in Different Types of Forest Soils in Japan. // *Soil Sci. Plant Nutr*. -1993.- V.39, P.485-495.

Tani M., Shida K. S., Tsutsuki K., Kondo R. Determination of Water-Soluble Low-molecular-Weight Organic Acids in Soils by Ion Chromatography. // *Soil Sci. Plant Nutr*.- 2001.-V.47.-No.2.-P. 387-397.

Tani M.and Higashi T. Vertical Distribution of Low-molecular-Weight Aliphatic Carboxylic Acids in Some Forest Soils of Japan. //*Eur. J. Soil Sci*. -1999.-V.50.- P.217-226.

Tipping E. Cation binding by humic substances. – Cambridge: Cambridge University Press, 2005. – P. 444.

Titeux H., Brahy V., and Delvaux B. Cu(II) complexation capacity of litter leachates and soil solutions in acid forest soils. // *Proc. Extended Abstracts of the 5th International Conference on the Biogeochemistry of Trace Elements, Vienna, Austria*.- 1999.-V. 1.-P. 374.

Titova E.V., Kasimova L.V., Sorokin I.B., and Panov A.N. Efficiency of humic plant growth regulators from peat. // *Proceedings of the IV All-Russian Conference "Humic Substances in the Biosphere."* - St. Petersburg, 2007. - P. 528-535.

Town R.M. and Powell K.J. Ion-selective electrode potentiometric studies on the complexation of copper(II) by soil-derived humic and fulvic acids. // *Anal. Chim. Acta*. – 1993. – V. 279. – P. 221-233.

Town R.M., Filella M. Dispelling the myths: Is the existence of L1 and L2 ligands necessary to explain metal ion speciation in natural waters? // *Limnol. Oceanogr*. – 2000. - V. 45. - P. 1341-1357.

Tulonen, T., Salonen, K. and Arvola, L. Effects of different molecular weight fractions of dissolved organic matter on the growth of bacteria, algae and protozoa from a highly humic lake. // *Hydrobiologia*.- 1992.- V.229.- P.239-252.

Turkmen O., Dursun A., Turan M., and Erdinc C. Calcium and humic acid affect seed germination, growth, and nutrient content of tomato (*Lycopersicon esculentum* L.) seedlings under saline soil conditions // *Acta Agric. Scand., Sect. B, Soil and Plant Sci.* – 2004. – V. 54. – P. 168-174.

Tyurin I.V. Soil Organic Matter and Its Role in Fertility. - Moscow: Akad. Nauk SSSR, 1965. - 108 p.

Tyurin I.V. Study of podzol formation. // *Pochvovedenie*.-1944.- No. 3.- P. 23-37.

Ushakova G. I. Specific Features of Litter Development and Transformation in Forest Biogeocenoses of the Khibiny Mountains. // Eur. Soil Sci.- 1999.- V.32.-No.12.-P.1305-1311.

Vakhmistrov D.B., Zverkova O.A., Debeets S.Yu., and Mishustina N.E. Humic acids: relationship between the surface activity and the stimulation of plant growth // *Dokl. AN SSSR*. -1987. - V. 293. - P. 1277-1280.

Van de Venter H.A., Furter M., Dekker J., and Cronje I.J. Stimulation of seedling root growth by coal-derived sodium humate // Plant and Soil. – 1991. – V. 138. – P. 17-21.

Van Hees P.A.W., Andersson A.-M.T., Lundström U.S. Separation of organic low molecular weight aluminium complexes in solution by liquid chromatography // *Chemosphere*.- 1996. - V. 33. - P. 1951-1966.

Van Hees P.A.W., Lundstrom U.S., Giesler R. Low molecular weight acids and their Al complexes in soil solution - composition, distribution and seasonal variation in three podzolized soils. // *Geoderma*.- 2000.- V. 94.- P.173-200.

Varshal G. M., Intskirveli L. N., Sirotkina I. S., et al. Association of Fulvic Acids in Water Solutions. // *Geokhimiya*. – 1975.-No. 1.- P.1581-1584.

Varshal G.M., Koshcheeva I.Ya., Sirotkina I.S., et al. Study of organic matter in surface waters and its interaction with metal ions. //*Geokhimiya*.-1979.- No. 4.- P. 598-607.

Varshal G.M., Velyukhanova T.K., and Koshcheeva I.Ya. Geochemical role of humic substances in the migration of elements. // *Humic Substances in the Biosphere*.- Moscow: Nauka, 1993. P.97-117.

Varshal G.M., Velyukhanova T.K., Koshcheeva I.Ya., Dorofeeva V.A., Buachidze N.S., Kasimova O.G., Makharadze G.A. The investigation of species of elements in surface water // *Zhurn. Anal. Khim.* – 1983. – V. 38. – P. 1590-1599.

Vaughan D., Malcolm R.E. Influence of Humic Substances on Growth and Physiological Processes // Soil Organic Matter and Biological Activity. / Ed. by D. Vaughan et al. - Dordrecht: Martinus Nijhoff. – 1985. - P. 37-75.

Vermeer A.W.P and Koopal L.K. Adsorption of humic acids to mineral particles. 2. Polydispersity effects with polyelectrolyte adsorption // *Langmuir.* – 1998. – V. 14. – P. 4210-4216.

Vernadskii V.I. History of Natural Waters. Part 1.- Leningrad: Chimteoret, 1936. - No. 3. -p.128.

Vincler P., Lacatov B., Mady Gy., Meisel J., and Mohos B. Infrared and EPR spectra of peats, peat humic acids and metal humats // *Humus Planta.* - 1971. – No. 5. - P. 301-304.

Vinogradov A.P. Geochemistry of Rare and Dispersed Chemical Elements in Soils.- Moscow: Akad. Nauk SSSR, 1957.-p.210.

Visser S.A. Effects of humic substances on plant growth. // Humic substances: Effect on soil and plants. - Reda, 1986. – P. 89-135.

Vorob'eva L.A. Theory and Practice of the Chemical Analysis of Soils. - Moscow: GEOS, 2006. - 400 p.

Wang Q., Li Z., Cheng S. Wu Z. Influence of humic acids on the accumulation of copper and cadmium in Vallisneria spiralis L. from sediment // Environ. Earth Sci. – 2010. - V. 61. - P. 1207-1213.

Wang S.C.T., Yang Tze-Ken, and Chuang Tzc-Tang. Soil Phenolic Acids as Plant Growth Inhibitors. //*J. Soil Sci.* -1967. -V.103.-No.4.-P. 67-72.

Wilson M. A., Collin P. J., and Tale K. P. ^{1}H Nuclear Magnetic Resonance Study of Soil Humic Acid. // *Soil Sci.*-1983.-No.24. P. 297-304.

Wu F., Tanoue E. Isolation and partial characterization of dissolved copper-complexing ligands in streamwaters // *Environ. Sci. Technol.* - 2001. - V. 35. - P. 3646.

Yashin I. M. Water-soluble organic substances of taiga soils and their ecological functions. Doctoral Dissertation in Biology. - Moscow, 1993.- 280 p.

Yin S., Sorom L., Jonson J. A., and Tyler G. Low-Molecular Acids (LOAC) in the Rhizosphere Soil Solution of Beech *(Fagus silvatica)* Forest Cambisols Determined by Ion Chromatography Using Supported Liquid Membrane Enrichment. // *Soil Biol. Biogeochem.* -1996.-V. 28.-No.9. P.1163-1169.

Yonebayashi K., Hattori T. Chemical and biological studies on environmental humic acids. II. ^{1}H-NMR and IR spectra of humic acids // *Soil Sci. Plant Nutr.* – 1989. V. 35. - P. 383-392.

Zaitseva T. F., Klevenskaya I. L., and Morozkov V. K. Low-Molecular Organic Acids in the Barguzin Hollow Soils. // *Pochvovedenie*.- 1987.- No. 3.- P.35-42.

Zhorobekova Sh.Zh., Mal'tseva G.M., and Kydralieva K.A. Features of complexation of humic acids with metal ions. // *Humic substances in the biosphere. Biological Sciences. Nauch. Dokl. Vyssh. Shkoly.* – 1991. – No. 10.

Zhuravleva E.G. Species and mobility of copper in soddy-podzolic soil. // *Proc. Dokuchaev Soil Science Institute, Series Soil Chemistry: Species and Methods of Determining Macro- and Microelements*.- Moscow, 1978.-P.34-48.

INDEX

C

D

H

I

J

K

L

M

N

O

P

Q

R

S

T

U

V

W

Y

Z